Springer Monographs in Mathematics

Springer
Berlin
Heidelberg
New York
Barcelona
Hong Kong
London
Milan
Paris
Tokyo

Helmut Koch

Galois Theory of p-Extensions

Springer

Helmut Koch
Humboldt-Universität Berlin
Institut für Mathematik
Unter den Linden 6
10117 Berlin
Gemany

Translator
Franz Lemmermeyer
California State University San Marcos
Department of Mathematics
333 South Twin Oaks Valley Rd.
92096-0001 San Marcos, CA
USA
e-mail: franzl@csusm.edu

Catalog-in-Publication Data applied for

Die Deutsche Bibliothek - CIP-Einheitsaufnahme

Koch, Helmut: Galois theory of p-extensions / Helmut Koch. - Berlin ; Heidelberg ; New York ; Barcelona ; Hong Kong ; London ; Milan ; Paris ; Tokyo : Springer, 2002
ISBN 3-540-43629-4

Mathematics Subject Classification (2000): 11R34, 11-02

ISBN 3-540-43629-4 Springer-Verlag Berlin Heidelberg New York

Springer-Verlag Berlin Heidelberg New York
a member of BertelsmannSpringer Science+Business Media GmbH

http://www.springer.de

Printed in Germany

Typeset in TEX by the Translator
Cover production: Erich Kirchner, Heidelberg

SPIN 10868507 44/3142db - 5 4 3 2 1 0 – Printed on acid-free paper

Foreword

This book deals with a quite young area of algebraic number theory: the algebraic theory of p-extensions, which was developed in the last 25 years and has now reached a degree of completion that makes a systematic presentation highly desirable.

This area of arithmetic deals with the theory of finite extensions of fields of arithmetic type. These are $\mathfrak{p}$-adic number fields, fields of formal power series with finite fields of constants, algebraic number fields and algebraic function fields of one variable with finite fields of constants. The main goal is obtaining information beyond what is provided by classical class field theory, which – as is well known – describes the extensions with commutative Galois group. The commutativity of the Galois group is very essential here. Class field theory is thereby closely connected with a wide circle of mathematical ideas ranging from the theory of radical extensions (nowadays called Kummer theory) to topological duality theorems, the theory of abelian and harmonic integrals and Picard varieties. The group-theoretical foundation of all these questions is the Pontryagin duality of abelian groups and their character groups. It is this part of mathematics that A. Weil has called "abelian mathematics".

As is well known, when Hilbert started building class field theory, he was led by the analogy between algebraic number fields and function fields, i.e., the field of meromorphic functions on compact Riemann surfaces. From this point of view, any "nonabelian" generalization of class field theory has to correspond to the investigation of the fundamental group of Riemann surfaces.

It has turned out that building a theory extending beyond the frame of class field theory is possible for extensions whose Galois group is nilpotent (or, what amounts to the same, a p-group). As in the study of the fundamental group in topology, it emerged that the groups of interest here have a natural presentation in terms of generators and relations, which clarifies essential parts of their structure. These investigations are contained in this book. Although the theory presented here is far from dealing with the general case of finite groups, it leads to the solution of many number-theoretic problems. For example, the theory says a lot about the structure of *all* extensions of a $\mathfrak{p}$-adic number field or of the field of formal power series over a finite field of constants. This theory also leads to the solution of the class field tower problem and of the problem connected with it, the growth of minimal discriminants of algebraic number fields.

At the same time I would like to stress that we are dealing here with a theory that is not yet complete. On the contrary, several extremely interesting problems remain unsolved. Examples of such questions are the following: a concrete description of the Galois group of the algebraic closure of a $\mathfrak{p}$-adic number field (e.g. including the ramification groups), the existence of extension fields of the rational number field with given (nonsolvable) Galois group, and the integrality of L-series for Artin characters different from the trivial character.

I am convinced that this book is of interest to a large circle of mathematicians. On the one hand, it is accessible to nonexperts and leads the reader quickly into a new area of investigation that contains a wealth of problems awaiting solution. On the other hand, the book contains almost all the basic material pertaining to this area as well as many new results by the author that will be of interest to the experts. The author's papers are among the most interesting results in this direction and have contributed essentially to the reputation that this area of mathematics has nowadays.

I hope that this book by H. Koch will stimulate the further development of this direction in algebraic number theory.

Moscow, December 1969 I.R. Shafarevich

Preface for the English Translation

There are a number of reasons why an English translation of this book appears more than 30 years after its publication.

1. It is very hard to find a copy of this book even from used book stores. On the other hand, it is still in demand since the subject has not been included in textbooks.

2. This book demonstrates that the cohomology of groups is very useful for studying Galois theory of number fields; at the same time, it offers a down to earth introduction to the cohomological method. Thus it may be read in parallel to Serre's classical lectures *Cohomologie Galoisienne* first published in the Springer Lecture Notes Series in 1964, and whose English translation appeared in 2000.

3. Another book published in 2000 is the encyclopedic [NSW] by J. Neukirch, A. Schmidt and K. Wingberg, where the results presented in this book are discussed only marginally. For example, the authors refer to the German original of this book for the improved version of the Golod-Shafarevich inequality. On the other hand, [NSW] contains new results on the Galois theory of p-extensions, in particular in connection with maximal p-extensions unramified outside of a set S of primes containing the primes dividing p. The present book, on the other hand, concentrates on extensions in which primes dividing p may be unramified.

Berlin, autumn 2001 H. Koch

Note by the Translator

It is my pleasure to thank Linda Holt for her comments, Farshid Hajir for suggesting the translation in the first place and for his contribution to the postscript, Wayne Aitken for corrections and explanations, Richard Hill, Hendrik Lenstra, Akito Nomura, Peter Pleasant and an unknown referee for lists of typos and errors, and Helmut Koch for his help in preparing this translation.

San Marcos, spring 2002 F. Lemmermeyer

Preface of the German Edition

The main objective of this book is a consistent presentation of results by I.R. Shafarevich, A. Fröhlich, A. Brumer and the author on the Galois theory of p-extensions on the basis of Galois cohomology. In order to make these results accessible to a wider circle of mathematicians with algebraic interests, prerequisites are restricted to knowledge of basic facts from algebra, group theory, and algebraic number theory as they are readily available in standard textbooks.

The first seven chapters of this book deal with cohomology of profinite groups, in particular of pro-p-groups. It goes without saying that the first chapter of Serre's Cohomologie Galoisienne served as a blueprint. Some chapters of this part may be seen as a commentary on Serre's lectures. The rest of the book deals with field theory. The theorems of class field theory that we need are stated explicitly and may be accepted axiomatically by the readers. There exist, by the way, at least two easily accessible presentations of class field theory in the style and extent in which we will need it, namely J.W.S. Cassels and A. Fröhlich [8] as well as J. Neukirch [46].

Some of the results of this book I found during a one-year stay in 1967/68 as a researcher at the Steklov Institute for Mathematics of the Academy of Science of the USSR. I would like to express my heartfelt gratitude to the Steklov Institute and in particular to Prof. I.R. Shafarevich for the invitation. I.R. Shafarevich's influence on this book is much larger, though: it goes back to my first stay in Moscow in 1960/61 as a student, which was followed by numerous inspiring and stimulating conversations.

I would like to thank the editors, in particular Prof. Dr. H. Reichardt, for including this book into the series "Mathematische Monographien". Moreover I thank Prof. Dr. H. Reichardt and the members of the number theory research group of the Institute for Pure Mathematics at the German Academy of Science in Berlin, namely Dr. O. Neumann, W. Thor, H. Pieper, and W. Zink, who have read parts of the manuscript and suggested improvements and corrections. I thank the VEB Deutscher Verlag der Wissenschaften, in particular the copy editors L. Boll and Miss E. Arndt, for their exemplary and expert work on the manuscript, which has contributed essentially to the success of this publication.

Berlin, autumn 1969 H. Koch

Contents

Introduction

Once the framework of Galois theory has been completed with the main theorem, the principal problem of the theory is the question: what are the possible normal extensions of a fixed base field k with given Galois group G. This problem is called the *Inverse Problem of Galois Theory.*

The solution of the inverse problem is connected intimately with properties of the base field k. In the simplest nontrivial example of the base field $k = \mathbb{R}$ of real numbers, the Fundamental Theorem of algebra yields the solution: If G is cyclic of order 2, then there is an extension with Galois group G, the field $\mathbb{C}$ of complex numbers. Groups G of higher order do not possess a representation as a Galois group. Another example where a complete solution of the problem is known are the function fields of one variable over algebraically closed fields of constants of characteristic 0 (cf. I.R. Shafarevich [57]).

The situation is completely different in the classical case where the base field k is an algebraic number field. Here we do not know whether an extension with an arbitrary group G exists. The most far-reaching result in this direction is the theorem of I.R. Shafarevich [56], which says that such an extension exists if G is solvable. A description of all possible extensions with given Galois group is provided by class field theory for the case of abelian groups.

In the present book we consider the larger class of nilpotent groups G, or, what amounts to the same thing by Galois theory, the class of p-groups for arbitrary primes p. The possible base fields are all fields accessible to class field theory, i.e. the global fields – algebraic number fields and function fields of one variable over a finite field of constants – and the local fields – $\mathfrak{p}$-adic number fields and the fields of power series in one variable over a finite field of constants.

First we want to make the formulation of the problem more precise. During the last 20 years, the idea of considering extensions with certain maximality properties turned out to be very fertile for Galois theory. The Galois group of such a (possibly infinite) extension is a profinite group in general, and, if we restrict ourselves to p-extensions, a pro-p group "which isn't very far from being free"; this suggests that a presentation of the group in terms of generators and relations may be adequate. More exactly, given a base field

k and a set S of places of k, we consider the compositum k_S (inside some fixed separable closure of k) of all normal extensions of k whose degree is a power of p and which are ramified only at places in S. This extension k_S is called the maximal p-extension of k unramified outside S. Each normal finite extension of k whose Galois group is a p-group is contained in a field k_S for some finite S. In the sense of Galois theory the inverse problem for p-groups is therefore equivalent with the description of the Galois group G_S of the extensions k_S/k for all finite S.

The main emphasis of this book is on the groups G_S for global base fields k. The p-extensions of $\mathfrak{p}$-adic number fields are treated only to the extent that is necessary for the transition to global fields. There are two reasons for this: on one hand, the inverse problem for such fields has been attacked successfully (see K. Iwasawa [25], Z.I. Borevich [2], H. Koch [29, 31], A.V. Yakovlev [64]), and it is to be expected that a complete solution is just around the corner. On the other hand there is a comprehensive presentation of the Galois theory of maximal p-extensions of $\mathfrak{p}$-adic number fields by J. Labute [41]. The corresponding theory for fields of power series over finite fields of constants is considerably simpler and will be treated completely.

The Galois group of the maximal p-extension of a local field $k_\mathfrak{p}$ has one generating relation or is free according as $k_\mathfrak{p}$ contains the pth roots of unity or not. If $\mathfrak{p} \in S$, then this relation can be associated to a relation in G_S. The question whether a generating system of relations can be obtained this way was discussed by Koch [32] and is the main motivation behind this book.

For algebraic number fields, the description of G_S is connected closely to the class field tower problem. This problem led Shafarevich [58] to consider the groups G_S.

With the help of structure theorems on G_S it is possible to obtain results on the p-component of the class group of abelian extensions of a base field k. The results presented in Chapter 12 of this book generalize theorems by A. Fröhlich [16, 17].

An essential result in the Galois Theory of p-extensions, the theorem of Scholz [50], Reichardt [48] and Shafarevich [55] on the existence of normal extensions of an algebraic number field with a given p-group as Galois group, is not discussed in this book. A proof of this result within the Galois cohomology on which this book is based has not yet been given.

1. Profinite Groups

Galois groups of infinite normal extensions are profinite groups. For this reason we shall start by discussing the general properties of these groups. For the basic material in the theory of topological groups we refer to L.S. Pontryagin [47]. A subgroup of a topological group will always be assumed to be a closed subgroup.

1.1 Projective Limits of Groups and Rings

Let I be a directed set, i.e., a partially ordered set such that for all $i, j \in I$ there exists a $k \in I$ with $i \leq k$ and $j \leq k$. We can look at I as a category whose objects are the elements of I, and whose sets of morphisms $\mathrm{Hom}(i, j)$, $i, j \in I$, consist of one element if $i \leq j$ and are empty otherwise.

Definition 1.1. A *projective system* $P = \{I, G_i, \varphi_i^j\}$ of compact groups (rings) is a contravariant functor from I into the category $\mathfrak{K}$ of compact groups (rings), where to $i \in I$ we associate the group (ring) G_i, and where φ_i^j is the morphism $G_j \longrightarrow G_i$ associated to $\mathrm{Hom}(i, j)$ for $i \leq j$. We will often write $\{G_i \mid i \in I\}$ for $\{I, G_i, \varphi_i^j\}$.

Example 1.2. Let $I = \mathbb{N}$ be the set of natural numbers (in their natural order), $\{G_i \mid i \in \mathbb{N}\}$ a family of compact groups, and let φ_i^{i+1} be arbitrary morphisms $G_{i+1} \longrightarrow G_i$. Whenever $i \leq j$, define

$$\varphi_i^j = \varphi_i^{i+1} \varphi_{i+1}^{i+2} \cdots \varphi_{j-1}^j.$$

Then $\{\mathbb{N}, G_i, \varphi_i^j\}$ is a projective system.

Definition 1.3. For every $G \in \mathfrak{K}$ we define the trivial projective system P_G assigning the group (ring) G to every $i \in I$ and the identity map to every $i \leq j$.

To each morphism $\varphi : G' \longrightarrow G$ for $G, G' \in \mathfrak{K}$ there is a natural transformation $P_{G'} \longrightarrow P_G$ that we will also denote by φ.

Definition 1.4. A pair (G, Φ) consisting of a group (ring) $G \in \mathfrak{K}$ and a natural transformation Φ from P_G to the projective system $P = \{I, G_i, \varphi_i^j\}$

is called a *projective limit* of P if, for every $G' \in \mathfrak{K}$ and every natural transformation $\Phi' : P_{G'} \longrightarrow P$, there exists a unique morphism $\varphi' : G' \longrightarrow G$ such that the diagram

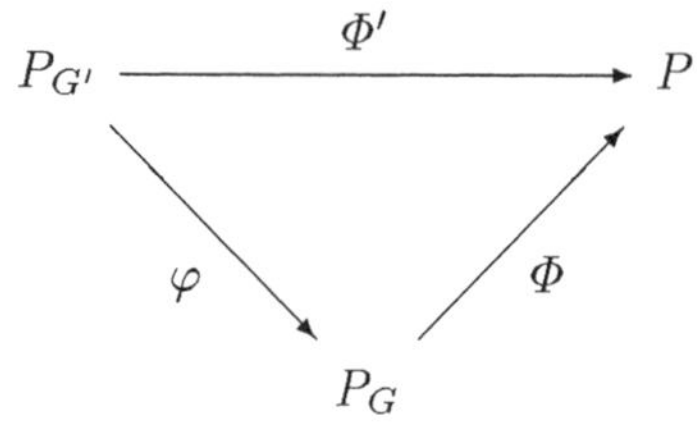

commutes.

A natural transformation from P_G to P is given by a family $\{\varphi_i \mid i \in I\}$ of morphisms $\varphi_i : G \longrightarrow G_i$ such that the diagram

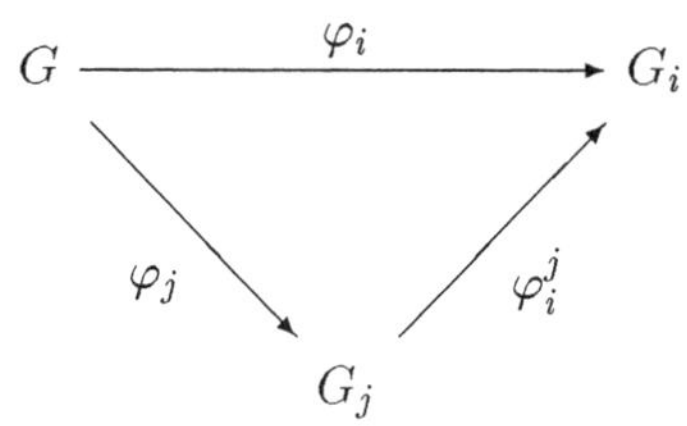

commutes for every $i \leq j$.

Theorem 1.5. *For every projective system $\{I, G_i, \varphi_i^j\}$, the projective limit in $\mathfrak{K}$ exists and is unique up to isomorphism in $\mathfrak{K}$.*

Proof. The uniqueness follows immediately from the definition. The projective limit $(G, \{\varphi_i \mid i \in I\})$ can be constructed as follows: G is the set of all elements $\prod_{i \in I} g_i$ in the direct product $\prod_{i \in I} G_i$ satisfying

$$g_i = \varphi_i^j g_j$$

for all $i, j \in I$ with $i \leq j$. The topology of G is induced by the compact topology of $\prod_{i \in I} G_i$. Since G is closed in $\prod_{i \in I} G_i$, it is compact. The morphism $\varphi_j : G \longrightarrow G_j$ is defined as the restriction of the projection $\prod_{i \in I} G_i \longrightarrow G_j$ to G. It is easily verified that the resulting topological group is the projective limit of the G_i. $\square$

We will denote the projective limit of the system $\{I, G_i, \varphi_i^j\}$ by $\varprojlim{}_{i \in I} G_i$ or simply by $\varprojlim G_i$. Projective systems (limits) are sometimes called inverse systems (limits).

Example 1.6. Let G be an arbitrary topological group (a topological ring) and I a set of normal subgroups (ideals) with finite index in G that is closed with respect to forming finite intersections. We make I into a directed set by putting

$$i \leq j \iff i \supseteq j \quad \text{for } i, j \in I.$$

For $i \leq j$ let φ_i^j be the projection

$$G/j \longrightarrow G/i.$$

Then $\{I, G/i, \varphi_i^j\}$ is a projective system whose limit we denote by G^I. The map

$$g \longmapsto \prod_{i \in I} gi \quad \left(\text{or } g \longmapsto \prod_{i \in I} (g+i) \quad \text{if } G \text{ is a ring}\right)$$

induces a morphism $\varphi_I : G \longrightarrow G^I$ whose kernel is the core of I, i.e., the intersection $\bigcap_{i \in I} i$.

If I consists of all normal subgroups (ideals) of finite index, then G^I is called the *profinite completion* of G; it will also be denoted by $\widehat{G}$.

Let p be a prime. If I consists of all normal subgroups (ideals) of p-power index in G, then G^I is called the *pro-p completion* of G.

Example 1.7. The pro-p completion of the ring $\mathbb{Z}$ of integers is isomorphic to the ring $\mathbb{Z}_p$ of p-adic integers. The profinite completion of $\mathbb{Z}$ is isomorphic to the direct product $\prod_p \mathbb{Z}_p$ over all primes p; this ring is often called the Prüfer ring $\widehat{\mathbb{Z}}$.

More generally, let $\mathcal{O}$ be a Dedekind domain and $\mathfrak{p}$ a prime ideal of $\mathcal{O}$ with finite residue class field. Then the completion of $\mathcal{O}$ with respect to the set I of all powers of $\mathfrak{p}$ is isomorphic to the ring $\mathcal{O}_\mathfrak{p}$ of integers in the completion of the quotient field K of $\mathcal{O}$ with respect to the valuation of K induced by $\mathfrak{p}$.

Assume that the residue class fields $\mathcal{O}/\mathfrak{p}$ are finite for all prime ideals $\mathfrak{p}$ of $\mathcal{O}$. Then the profinite completion of $\mathcal{O}$ is isomorphic to the direct product $\prod \mathcal{O}_\mathfrak{p}$ over all prime ideals $\mathfrak{p}$ of $\mathcal{O}$.

Let $P = \{I, G_i, \varphi_i^j\}$ and $Q = \{J, H_i, \psi_i^j\}$ be projective systems. A morphism $\varphi : P \longrightarrow Q$ consists of a morphism $\Phi : J \longrightarrow I$ of directed sets and of morphisms $\psi_i : G_{\Phi(i)} \longrightarrow H_i$ for every $i \in J$ such that the diagram

$$\begin{array}{ccc} G_{\Phi(i)} & \xrightarrow[\psi_i]{} & H_i \\ \uparrow {\scriptstyle \varphi_{\Phi(i)}^{\Phi(j)}} & & \uparrow {\scriptstyle \psi_i^j} \\ G_{\Phi(j)} & \xrightarrow[\psi_j]{} & H_j \end{array}$$

commutes for all $i \leq j$.

To each such morphism φ we can associate a morphism $\varphi' : \varprojlim G_i \longrightarrow \varprojlim H_i$ as follows:

$$\varphi'\left(\prod_{i \in I} g_i\right) = \prod_{j \in J} \psi_j(g_{\Phi(j)}).$$

Definition 1.8. Let I be a directed set. A subset J of I is called *cofinal* in I if for every $i \in I$ there is a $j \in J$ such that $i \leq j$.

For example, every infinite subset of $\mathbb{N}$ is cofinal in $\mathbb{N}$.

Now let $P = \{I, G_i, \varphi_i^j\}$ be a projective system. Restricting to $J \subseteq I$ gives us a projective system $Q = \{J, G_i, \varphi_i^j\}$. The inclusion $J \hookrightarrow I$, together with the identities $G_i \longrightarrow G_i$, induces a morphism $\varphi : P \longrightarrow Q$. The corresponding morphism $\varphi' : \varprojlim_{i \in I} G_i \longrightarrow \varprojlim_{i \in J} G_i$ is an isomorphism. In general, we shall identify $\varprojlim_{i \in I} G_i$ and $\varprojlim_{i \in J} G_i$.

The preceding discussion makes it clear what the category $\mathfrak{K}_I$ of projective systems over I is (we take the identity as our morphism $\Phi : I \longrightarrow I$). The functor $\varprojlim_{i \in I} : \mathfrak{K}_I \longrightarrow \mathfrak{K}$ is exact; in other words:

Theorem 1.9. *Let $P = \{I, F_i, \theta_i^j\}$, $Q = \{I, G_i, \varphi_i^j\}$ and $R = \{I, H_i, \psi_i^j\}$ be projective systems, let $\theta = \{\theta_i\}$ be a morphism $P \longrightarrow Q$ and $\psi = \{\psi_i\}$ be a morphism $Q \longrightarrow R$. For all $i \in I$, assume that*

$$1 \longrightarrow F_i \xrightarrow{\theta_i} G_i \xrightarrow{\psi_i} H_i \longrightarrow 1$$

is an exact sequence. Then

$$1 \longrightarrow \varprojlim F_i \xrightarrow{\theta'} \varprojlim G_i \xrightarrow{\psi'} \varprojlim H_i \longrightarrow 1$$

is also an exact sequence.

Proof. The morphism ψ' is surjective: let $\prod_{i \in I} h_i$ be an element in $\varprojlim H_i$. The sets

$$A_j = \left\{ \prod_{i \in I} g_i \;\middle|\; g_i \in G_i, \psi_j g_j = h_j, \varphi_i^j g_j = g_i \text{ for all } i \leq j \right\}$$

are closed, nonempty, and have the finite intersection property: the intersection of finitely many A_j is nonempty. Since $\prod_{i \in I} G_i$ is compact, $\bigcap_{j \in I} A_j$ is nonempty. Each element in $\bigcap_{j \in I} A_j$ belongs to $\varprojlim G_i$ and is mapped by ψ' to $\prod_{i \in I} h_i$.

The fact that $\ker \psi' \subseteq \operatorname{im} \theta'$ is proved similarly. The other claim follows immediately from the definitions. □

The following result is proved in an analogous manner:

Theorem 1.10. *Let $\varprojlim G_i = (G, \{\varphi_i \mid i \in I\})$ be the limit of the projective system $\{I, G_i, \varphi_i^j\}$ of groups, j an index in I, and $g_j \in G_j$. If $g_j \in \operatorname{im} \varphi_i^j$ for all $i \in I$ such that $i \geq j$, then $g_j \in \operatorname{im} \varphi_j$.*

1.2 Profinite Groups

Definition 1.11. A *profinite group* is a topological group that can be realized as a projective limit of finite groups.

Example 1.12. The groups G^I discussed in Example 1.6 are profinite groups; in particular, $\mathbb{Z}_p$ is a profinite group.

Theorem 1.13. *Let* $(G, \{\varphi_i \mid i \in I\})$ *be the projective limit of the projective system* $\{I, G_i, \varphi_i^j\}$ *of finite groups* G_i. *Then* $\{\ker \varphi_i \mid i \in I\}$ *is a neighborhood basis at the unit element* $1 \in G$.

Proof. Theorem 1.13 follows immediately from the definition of the topology in G. □

The following result gives an intrinsic characterization of profinite groups.

Theorem 1.14. *Let G be a compact Hausdorff group. Then the following assertions are equivalent:*

(i) *G is profinite;*

(ii) *G is totally disconnected;*

(iii) *There is a set $\mathfrak{U}$ of open normal subgroups of G that form a neighborhood basis at the unit element in G.*

Proof. By Theorem 1.13, (i) implies (iii). Assume conversely that (iii) holds. Then $\{G/U \mid U \in \mathfrak{U}\}$ forms a projective system of finite groups (see Example 1.6). Now we apply Theorem 1.9 to the projective systems $\{U \mid U \in \mathfrak{U}\}$, $\{G \mid U \in \mathfrak{U}\}$ and $\{G/U \mid U \in \mathfrak{U}\}$ with the morphisms $\varphi_U : U \longrightarrow G$ and $\psi_U : G \longrightarrow G/U$ for $U \in \mathfrak{U}$. Since

$$\varprojlim{}_{U \in \mathfrak{U}} U = \{1\},$$

we find

$$G = \varprojlim{}_{U \in \mathfrak{U}} G/U,$$

hence (iii) implies (i).

The condition (ii) is equivalent to demanding that the intersection of all open and closed neighborhoods of 1 equals $\{1\}$. Thus (ii) follows immediately from (iii). According to Pontryagin[1] [47, III, Thm. 17],(ii) implies (iii). □

In the following, $\mathfrak{U}_G$ will denote the system of all normal subgroups of finite index in G.

The preceding discussion easily implies the following

[1] See also Ribes [Ri, Lemma 2.4], or Wilson [Wil, Prop. 0.3.3].

Theorem 1.15. *Direct products and projective limits of profinite groups are profinite.*

More generally it can be shown that projective limits exist in the category of profinite groups (morphisms are the continuous group homomorphisms) whenever the index system I is a small category; the analog of Theorem 1.9, however, does not always hold. Let us also mention the notion of *fibered products*: to every diagram

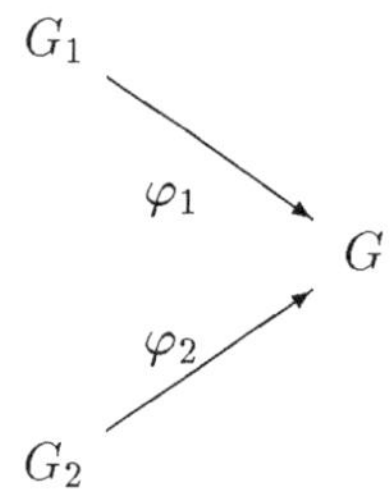

of profinite groups there is (up to isomorphism) a unique profinite group $G_1 \times_G G_2$ with morphisms to G_1 and G_2 such that the diagram

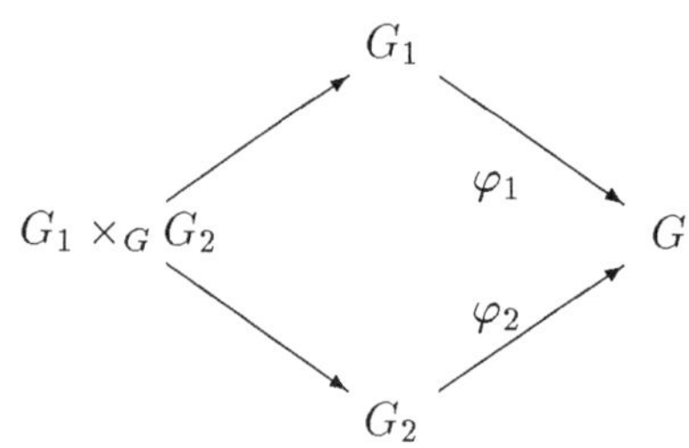

commutes, and such that for every commutative diagram

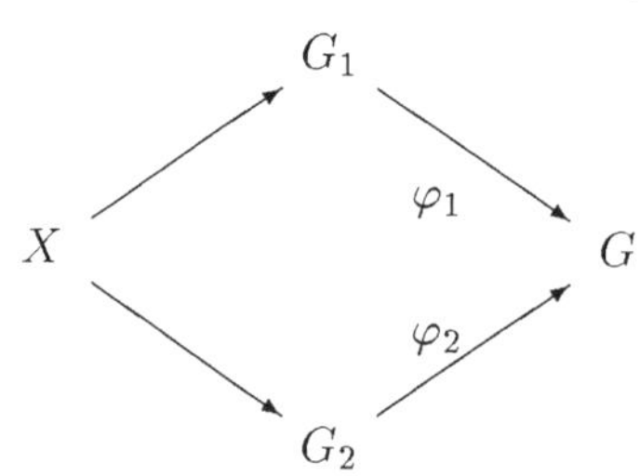

there is a unique morphism $X \longrightarrow G_1 \times_G G_2$ that makes the diagram

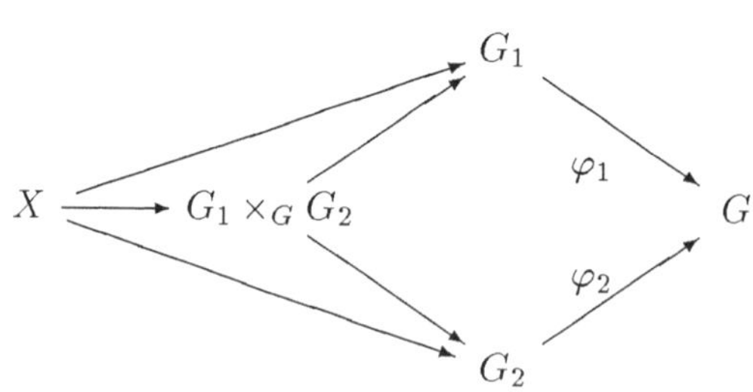

commutative.

The group $G_1 \times_G G_2$ consists of all pairs $(g_1, g_2) \in G_1 \times G_2$ such that $\varphi_1 g_1 = \varphi_2 g_2$.

1.3 Subgroups and Quotient Groups

A subgroup V of a profinite group G has $\{V \cap U \mid U \in \mathfrak{U}_G\}$ as a neighborhood basis at 1, hence is profinite by Theorem 1.14 and the first isomorphism theorem.

A quotient group of G modulo a normal subgroup N has $\{UN/N \mid U \in \mathfrak{U}_G\}$ as a neighborhood basis at 1, hence is profinite by Theorem 1.14 and the second isomorphism theorem.

The following theorem guarantees the existence of a *continuous section*, that is, a continuous system of representatives of cosets with respect to a subgroup.

Theorem 1.16. *Let H be a subgroup of the profinite group G. Then there is a continuous section $\sigma : G/H \longrightarrow G$ such that $\sigma(H) = 1$, that is, a continuous map $\sigma : G/H \longrightarrow G$ between topological spaces such that the composition*

$$G/H \xrightarrow{\ \sigma\ } G \longrightarrow G/H$$

is the identity map on G/H.

Before we prove Theorem 1.16, we formulate some auxiliary results.

Let $\mathfrak{V}$ be a filtered system of subgroups of G, and write $S = \bigcap_{V \in \mathfrak{V}} V$. Then $\{G/V : V \in \mathfrak{V}\}$ is a projective system of topological spaces. The considerations in §1.1 and §1.2 may be transferred to this situation almost word for word. In particular, the projective limit of $\{G/V : V \in \mathfrak{V}\}$ exists, and the natural continuous map $G/S \longrightarrow \varprojlim_{V \in \mathfrak{V}} G/V$ is a homeomorphism.

Lemma 1.17. *Let $K \subseteq H$ be subgroups of G such that H/K is finite. Then there exists a continuous section $\sigma : G/H \longrightarrow G/K$ such that $\sigma(H) = K$.*

Proof. $\{UK/K \mid U \in \mathfrak{U}_G\}$ is a neighborhood basis at $1 \in G/K$. Since H/K is finite, there is a $U \in \mathfrak{U}_G$ such that $UK/K \cap H/K = \{1\}$. This implies that the natural map

$$\varphi : UK/K \longrightarrow UH/H$$

is a homeomorphism. Let $g_1 = 1, g_2, \ldots, g_k$ be a complete system of representatives for G/UH. Then we set

$$\sigma(g_\nu uH) = g_\nu \varphi^{-1}(uH), \quad \nu = 1, \ldots, k;\ u \in U.$$

This σ does it. □

Proof of Theorem 1.16. Let $\mathcal{X}$ denote the set of all pairs (S,σ), where S is a subgroup of H and σ a continuous section $G/H \longrightarrow G/S$. We can partially order $\mathcal{X}$ by putting $(S,\sigma) \le (S',\sigma')$ if $S' \subseteq S$ and if σ is induced by σ'. Then $\mathcal{X}$ is an inductive set: in fact, if $\{(S_i,\sigma_i) \mid i \in I\}$ is a chain such that $\bigcap_{i\in I} S_i = S$, then the sections $\sigma_i : G/H \longrightarrow G/S_i$ induce a continuous map $\sigma' : G/H \longrightarrow \varprojlim G/S_i$ and therefore a continuous section $\sigma : G/H \longrightarrow G/S$. Thus (S,σ) is an upper bound for $\{(S_i,\sigma_i) \mid i \in I\}$. By Zorn's Lemma, there exists a maximal element (S,σ) in $\mathcal{X}$. We must have $S = \{1\}$: otherwise there would exist a proper subgroup of finite index in S, and (S,σ) would not be maximal by Lemma 1.17. This proves our claims. □

Example 1.18. All subgroups $\neq \{0\}$ of the additive group $(\mathbb{Z}_p,+)$ have finite index and are isomorphic to $(\mathbb{Z}_p,+)$. For every prime power p^n, there is a unique subgroup of index p^n.

1.4 Abelian Profinite Groups, Pontryagin's Duality Theory

Let G be an abelian profinite group and $\chi : G \longrightarrow \mathbb{R}/\mathbb{Z}$ a morphism of topological groups. Since G is compact, so is $\operatorname{im} \chi$, and $\operatorname{im} \chi \simeq G/\ker \chi$. This implies that $\operatorname{im} \chi$ is finite, hence cyclic, and therefore contained in $\mathbb{Q}/\mathbb{Z}$. Thus the group that is dual to G in the sense of Pontryagin is $G^* = \operatorname{Hom}(G,\mathbb{Q}/\mathbb{Z})$. The group G^* is a discrete torsion group.

More exactly, * is an exact contravariant functor from the category of abelian profinite groups to the category of abelian discrete torsion groups and vice versa, taking direct products of profinite groups corresponds to direct sums of discrete torsion groups.

Example 1.19. The dual group of the profinite completion (see Example 1.6) of $\mathbb{Z}$ is $\mathbb{Q}/\mathbb{Z}$.

Example 1.20. The dual notion of a projective limit of compact abelian groups is the injective limit of discrete abelian groups (see §1.7). Let $\{G_i \mid i \in I\}$ be a projective system of abelian profinite groups. Then

$$(\varprojlim G_i)^* = \varinjlim G_i{}^*.$$

1.5 Discrete Modules

Let G be a profinite group and A a unitary G-module. We make A into a topological space by giving it the discrete topology.

Definition 1.21. A is called a *discrete G-module* if the map $G \times A \longrightarrow A$ induced by the action of G on A is continuous.

Theorem 1.22. *Let G be a profinite group, A a unitary G-module. Then the following assertions are equivalent:*

(i) *A is a discrete G-module;*

(ii) *for all $a \in A$, the stabilizer $G_a = \{g \in G \mid ga = a\}$ is an open subgroup of G;*

(iii) *Let $\mathfrak{U}$ be a neighborhood basis at $1 \in G$ consisting of open normal subgroups of G. Then*

$$A = \bigcup_{U \in \mathfrak{U}} A^U \quad \textit{where} \quad A^U = \{a \in A \mid ua = a \textit{ for all } u \in U\}.$$

Proof. (i) implies (ii): let $a \in A$. The preimage of $\{a\}$ under the restriction of $G \times A \longrightarrow A$ to $G \times \{a\} \longrightarrow A$ is $G_a \times \{a\}$. Therefore, G_a is open.

(ii) implies (iii): let $a \in A$; then G_a contains some $U \in \mathfrak{U}$. Thus $a \in A^{G_a} \subseteq A^U$.

(iii) implies (i): Let $a, b \in A$ and $g \in G$ such that $ga = b$. By (iii), b is invariant under some $U \in \mathfrak{U}$. Thus $Ug \times \{a\}$ is an open neighborhood of (g, a) whose image under $G \times A \longrightarrow A$ is $\{b\}$. □

Theorem 1.22 implies immediately that submodules and quotients of discrete G-modules are again discrete G-modules.

Example 1.23. Let A be an abelian group and G a profinite group that acts trivially on A. Then A is a discrete G-module.

Example 1.24. Let $\mathbb{Q}_p$ be the field of rational p-adic numbers. Then

$$a(b + \mathbb{Z}_p) = ab + \mathbb{Z}_p \quad \text{for } a \in \mathbb{Z}_p, b \in \mathbb{Q}_p$$

defines a discrete $\mathbb{Z}_p$-module structure on $\mathbb{Q}_p/\mathbb{Z}_p$.

Let G be a profinite group, H a subgroup of G, and A a discrete H-module. We denote by $M_G^H(A)$ the set of all continuous maps $f : G \longrightarrow A$ such that

$$hf(x) = f(hx) \quad \text{for all } h \in H, x \in G.$$

The map f is continuous if and only if there exists an open normal subgroup N of G such that f only depends on the cosets of G modulo N. In fact, $\{f^{-1}(a) \mid a \in A\}$ is an open covering of G. Thus there exists a finite subcovering $\{f^{-1}(a_\nu) \mid \nu = 1, \ldots, s\}$. Now $f^{-1}(a_\nu)$ is the union of open sets of the form gV, $g \in G$, V open normal subgroup of G. Since $f^{-1}(a_\nu)$ is also closed, finitely many gV suffice for covering $f^{-1}(a_\nu)$. Now let U be an open normal subgroup of G contained in all V that occur in the finite coverings of $f^{-1}(a_1), \ldots, f^{-1}(a_s)$. Then U does it.

Now we transfer the addition from A to $M_G^H(A)$ and define a multiplication by elements $g \in G$ on $M_G^H(A)$:

$$(gf)(x) = f(xg) \quad \text{for } g \in G. \tag{1.1}$$

As f depends only on the cosets of G/U, so does gf, hence gf is continuous. It is easily checked that (1.1) makes $M_G^H(A)$ into a discrete G-module.

We put $M_G^{\{1\}}(A) = M_G(A)$. Then $M_G(A)$ is called an *induced module*.

Let A be a discrete G-module. Then we can view A as an H-module. For $a \in A$,

$$f_a(x) = xa, \quad x \in G,$$

defines an element $f_a \in M_G^H(A)$. The map $a \longmapsto f_a$ induces an injection $A \hookrightarrow M_G^H(A)$ of G-modules.

Example 1.25. Let X be a discrete abelian group. Define a G-module structure on $X[G] = \mathbb{Z}[G] \otimes_{\mathbb{Z}} X$ by

$$h(g \otimes x) = hg \otimes x \quad \text{for } g, h \in G, x \in X.$$

If G is a finite group, then $X[G] \simeq M_G(X)$, the isomorphism being induced by the map

$$f \longmapsto \sum_{g \in G} g \otimes f(g^{-1}).$$

1.6 The Category $\mathcal{C}$

In the following, by a G-module we will always mean a discrete G-module. Apart from the category $\mathcal{C}_G$ of all G-modules, the following category $\mathcal{C}$ will be important.

The objects of $\mathcal{C}$ are the pairs $[G, A]$, where G is a profinite group and A a G-module. A morphism $[G, A] \longrightarrow [H, B]$ is a pair $[\varphi, \psi]$, where $\varphi : H \longrightarrow G$ is a morphism of profinite groups and where $\psi : A \longrightarrow B$ is a homomorphism of abelian groups such that the following compatibility condition is satisfied: for all $h \in H$ and $a \in A$ we have

$$\psi(\varphi(h)a) = h\psi(a). \tag{1.2}$$

Example 1.26. Let A be a G-module and let H be a subgroup of G. The injection $\varphi : H \hookrightarrow G$ and the identity map $A \longrightarrow A$ induce a morphism $[G, A] \longrightarrow [H, A]$ called the *restriction*.

Now let H be a normal subgroup of G. The module A^H of all elements of A invariant under the action of H can be viewed as a G/H-module. The maps $\varphi : G \longrightarrow G/H$ and the injection $\psi : A^H \hookrightarrow A$ define a morphism $[G/H, A^H] \longrightarrow [G, A]$ called the *inflation*.

Example 1.27. Let H be an arbitrary subgroup of G and A a H-module. The injection $H \hookrightarrow G$ and the homomorphism $\psi : M_G^H(A) \longrightarrow A$ that maps each $f \in M_G^H(A)$ to $f(1)$ form a morphism $[G, M_G^H(A)] \longrightarrow [H, A]$.

1.7 Inductive Limits in $\mathcal{C}$

Let I be a directed set.

Definition 1.28. An *inductive system* $\{I, [G_i, A_i], [\varphi_i^j, \psi_i^j]\}$ in $\mathcal{C}$ is a covariant functor $I \longrightarrow \mathcal{C}$, where $i \in I$ is assigned the object $[G_i, A_i]$, and where $[\varphi_i^j, \psi_i^j]$ is the morphism $[G_i, A_i] \longrightarrow [G_j, A_j]$ assigned to $\mathrm{Hom}\,(i,j)$ for $i \leq j$. Here $\{I, G_i, \varphi_i^j\}$ is a projective system in the category of profinite groups.

In the special case where $G_i = \{1\}$ for all $i \in I$, we get an inductive system $\{I, A_i, \psi_i^j\}$ of abelian groups.

Example 1.29. Let A be a G-module and $\{A_i \mid i \in \mathbb{N}\}$ an ascending sequence of submodules. For $i \leq j$ let φ_i^j be the identity map on G and ψ_i^j the injection $A_i \hookrightarrow A_j$. Then $\{\mathbb{N}, [G, A_i], [\varphi_i^j, \psi_i^j]\}$ is an inductive system.

Example 1.30. Let $\{K_i \mid i \in \mathbb{N}\}$ be an ascending sequence of finite normal extensions K_i of a field K, φ_i^j for $i \leq j$ the projection of the Galois groups $\mathrm{Gal}\,(K_j/K) \longrightarrow \mathrm{Gal}\,(K_i/K)$, and ψ_i^j the injection $K_i^+ \hookrightarrow K_j^+$. Then

$$\{\mathbb{N}, [\mathrm{Gal}(K_i/K), K_i^+], [\varphi_i^j, \psi_i^j]\}$$

is an inductive system.

Definition 1.31. Let I be a directed set. We assign to every G-module A the trivial inductive system $D_{[G,A]} = \{I, [G, A], [\varphi_i^j, \psi_i^j]\}$, where $\varphi_i^j : G \longrightarrow G$ and $\psi_i^j : A \longrightarrow A$ are the identity maps. A morphism $[\varphi, \psi] : [G, A] \longrightarrow [G', A']$ induces in an obvious way a natural transformation $D_{[G,A]} \longrightarrow D_{[G',A']}$, which we will also denote by $[\varphi, \psi]$.

Definition 1.32. A G-module A and a natural transformation

$$\Phi : D = \{I, [G_i, A_i], [\varphi_i^j, \psi_i^j]\} \longrightarrow D_{[G,A]}$$

of inductive systems is called an *inductive limit* of D, if, for every G'-module A' and every natural transformation $\Phi' : D \longrightarrow D_{[G',A']}$, there is a unique morphism $[\varphi, \psi] : [G, A] \longrightarrow [G', A']$ such that the diagram

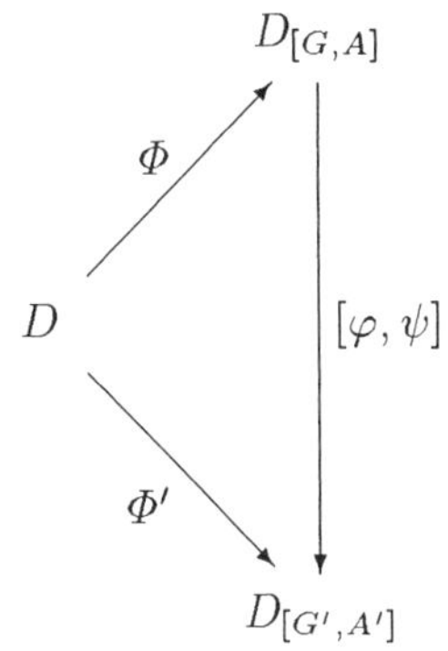

is commutative.

A natural transformation $D \longrightarrow D_{[G,A]}$ is determined by a family $\{[\varphi_i, \psi_i] \mid i \in I\}$ of morphisms $[G_i, A_i] \longrightarrow [G, A]$ such that for all $i \leq j$ the diagrams

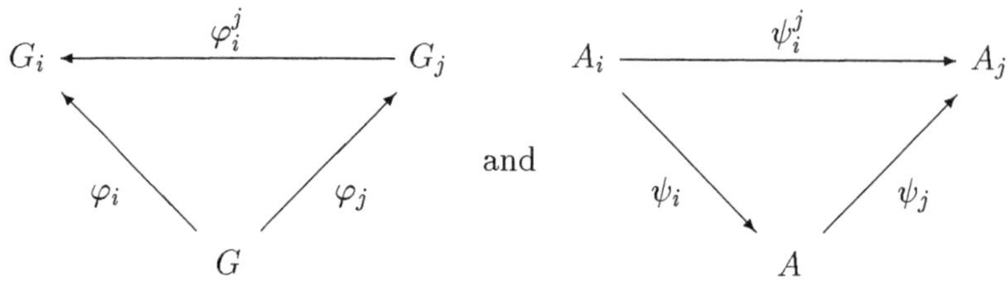

commute.

The inductive limit is unique up to isomorphism, and we have

Theorem 1.33. *In the category $\mathcal{C}$, inductive limits exist.*

Proof. Let $D = \{I, [G_i, A_i], [\varphi_i^j, \psi_i^j]\}$ be an inductive system. Then $\{I, G_i, \varphi_i^j\}$ is a projective system of profinite groups. In the following, we will use the construction of $\varprojlim G_i = G$ explained in the proof of Theorem 1.5.

The triple $\{I, A_i, \psi_i^j\}$ is an inductive system of abelian groups. The inductive limit $\varinjlim A_i = A$ of this system may be constructed as follows. Two elements $a_i \in A_i$, $a_j \in A_j$ are called equivalent $(a_i \sim a_j)$ if there is an index $k \in I$ with $i \leq k$, $j \leq k$ and

$$\psi_i^k(a_i) = \psi_j^k(a_j).$$

In this way we get an equivalence relation on the disjoint union of the A_i, $i \in I$. The equivalence classes with respect to $\sim$ form an abelian group, the inductive limit of $\{I, A_i, \psi_i^j\}$, in an obvious way. The corresponding maps $\psi_i : A_i \longrightarrow A$ map an element $a_i \in A_i$ to its class $\overline{a_i} \in A$.

We now define the action of elements $\prod_{i \in I} g_i \in G$ on $\overline{a_j} \in A$ by

$$\prod_{i \in I} g_i \overline{a_j} = \overline{g_j a_j}. \tag{1.3}$$

This definition does not depend on the choice of the representative a_j in its class: in fact, if $\overline{a_j} = \overline{a_{j'}}$, then there is an index $k \in I$ such that $j \leq k$, $j' \leq k$ and

$$\psi_j^k(a_j) = \psi_{j'}^k(a_{j'}).$$

But then (1.2), $g_j = \varphi_j^k g_k$ and $g_{j'} = \varphi_{j'}^k g_k$ imply that

$$\psi_{j'}^k(g_{j'} a_{j'}) = g_k \psi_{j'}^k(a_{j'}) = g_k \varphi_j^k(a_j) = \psi_j^k(g_j a_j).$$

It is easily checked that (1.3) defines a discrete G-module structure on A. Thus $[G, A]$ is the desired inductive limit, and the corresponding morphisms $[G_i, A_i] \longrightarrow [G, A]$ are given by the compatible pairs $[\varphi_i, \psi_i]$. □

The following result is easy to prove:

Theorem 1.34. *Let* $D_\nu = \{[G_i, A_i^{(\nu)}], i \in I\}$, $\nu = 1, 2, 3$, *be inductive systems in* $\mathcal{C}$, *and assume that*

$$0 \longrightarrow D_1 \longrightarrow D_2 \longrightarrow D_3 \longrightarrow 0$$

is an exact sequence. Then the induced sequence

$$0 \longrightarrow \varinjlim A_i^{(1)} \longrightarrow \varinjlim A_i^{(2)} \longrightarrow \varinjlim A_i^{(3)} \longrightarrow 0$$

is an exact sequence of $\varprojlim G_i$*-modules.*

2. Galois Theory of Infinite Algebraic Extensions

A Galois theory of a category $\mathfrak{K}$ is a contravariant functor of $\mathfrak{K}$ into a "simpler" category $\mathfrak{K}'$, where certain properties of the objects and morphisms of $\mathfrak{K}$ are reflected in $\mathfrak{K}'$.

In classical Galois theory, the objects of $\mathfrak{K}$ are normal finite extensions K/k (in the following, a normal extension will always mean a normal algebraic separable extension of fields), and the morphisms $K/k \longrightarrow K'/k'$ are the isomorphisms $K \longrightarrow K'$ that map k into k'. The category $\mathfrak{K}'$ is the category of finite groups, and the contravariant functor assigns to each K/k the group $\mathrm{Gal}\,(K/k)$ of automorphisms of K that leave k pointwise fixed, and to each morphism $K/k \longrightarrow K'/k'$ the corresponding projection $\mathrm{Gal}\,(K'/k') \longrightarrow \mathrm{Gal}\,(K/k)$.

In the transition from classical Galois theory to the Galois theory of infinite extensions, the category of finite groups is replaced by the category of profinite groups. Most results of the classical theory can be transferred without problems.

2.1 The Galois Group of Infinite Extensions

Let K/k be a (finite or infinite) normal extension.

Definition 2.1. The *Galois group* $\mathrm{Gal}\,(K/k)$ of K/k is defined as in the finite case as the group of all automorphisms of K leaving the elements of k fixed. We define a topology on $\mathrm{Gal}\,(K/k)$ via a neighborhood basis $\mathfrak{U}(K/k)$ at the unit element:

$$\mathfrak{U}(K/k) = \{\mathrm{Gal}\,(K/N) \mid N \in \mathcal{N}\},$$

where $\mathcal{N}$ is the set of all subfields of K that are normal and finite over k.

Theorem 2.2. *The group* $\mathrm{Gal}\,(K/k)$ *is profinite. More exactly,* $\mathrm{Gal}\,(K/k)$ *is isomorphic to the projective limit of the system* $\{\mathrm{Gal}\,(N/k) \mid N \in \mathcal{N}\}$.

Proof. The restriction $\mathrm{Gal}\,(K/k) \longrightarrow \mathrm{Gal}\,(N/k)$ induces a homomorphism $\varphi : \mathrm{Gal}\,(K/k) \longrightarrow \varprojlim_{N\in\mathcal{N}} \mathrm{Gal}\,(N/k)$. For every $\alpha \in K$ there is an $M \in \mathcal{N}$ such that $\alpha \in M$. We put

$$\psi\Big(\prod_{N\in\mathcal{N}} g_N\Big)(\alpha) = g_M(\alpha) \quad \text{for} \prod_{N\in\mathcal{N}} g_N \in \varprojlim_{N\in\mathcal{N}} \mathrm{Gal}\,(N/k).$$

This definition does not depend on the choice of M. It is easy to see that ψ is the inverse mapping of φ. Therefore, φ is an isomorphism of abstract groups. Since φ maps $\{\mathrm{Gal}\,(K/N)\,|\,N\in\mathcal{N}\}$, a neighborhood basis at $1\in\mathrm{Gal}\,(K/k)$, to the corresponding neighborhood basis in the projective limit $\varprojlim_{N\in\mathcal{N}}\mathrm{Gal}\,(N/k)$ (cf. Theorem 1.13), φ is even an isomorphism of topological groups. □

Theorem 2.3. *Let $\{K_i\,|\,i\in I\}$ be a family of normal extensions of k and put $K=\bigcup_{i\in I}K_i$. Assume that, for all $i,j\in I$ there is an $n\in I$ such that $K_i,K_j\subseteq K_n$. Then $\{\mathrm{Gal}\,(K_i/k)\,|\,i\in I\}$ is a projective system with limit $\varprojlim_{i\in I}\mathrm{Gal}\,(K_i/k)\simeq\mathrm{Gal}\,(K/k)$.*

Proof. The proof of Theorem 2.3 is analogous to the proof of Theorem 2.2. □

Example 2.4. Let p be a prime and put $k=\mathbb{Q}(\zeta_p)$ and $K=\bigcup_{\nu=1}^{\infty}\mathbb{Q}(\zeta_{p^\nu})$, where ζ_{p^ν} is a primitive p^ν-th root of unity. Then $\mathrm{Gal}\,(K/k)\simeq\mathbb{Z}/2\mathbb{Z}\times\mathbb{Z}_2$ if $p=2$, and $\mathrm{Gal}\,(K/k)\simeq\mathbb{Z}_p$ if $p>2$.

Theorem 2.5 (Lifts of Isomorphisms). *Let K/k be a normal extension with subextensions K_1/k and K_2/k, and let $\gamma: K_1\longrightarrow K_2$ be an isomorphism leaving the elements of k fixed. Then γ can be lifted to an automorphism of K/k.*

Proof. For every $M\in\mathcal{N}$ we define the subsets

$$B_M = \Big\{\prod_{N\in\mathcal{N}} g_N \,|\, g_M \text{ is a lift of } \gamma_{K_1\cap M} \text{ to } M,\ g_N = g_M|_N \text{ for} N\subseteq M\Big\}$$

of $\prod_{N\in\mathcal{N}}\mathrm{Gal}\,(N/k)$. From finite Galois theory we know that $\{B_M\,|\,M\in\mathcal{N}\}$ is a system of closed subsets of the compact set $\prod_{N\in\mathcal{N}}\mathrm{Gal}\,(N/k)$ having the finite intersection property. The intersection B of $\{B_M\,|\,M\in\mathcal{N}\}$ is therefore nonempty, and every element of B is a lift of γ to K by Theorem 2.2. □

Theorem 2.6. *If K/k is a normal extension and M/k a finite subextension, then $(\mathrm{Gal}\,(K/k):\mathrm{Gal}\,(K/M))=(M:k)$.*

Proof. Let $(M:k)=n$, and let $\gamma_1,\ldots,\gamma_n$ denote the isomorphisms $M\longrightarrow K$. By Theorem 2.5, γ_ν can be lifted to K. If $\overline{\gamma}_\nu$ is a lift of γ_ν, then $\{\overline{\gamma}_1,\ldots,\overline{\gamma}_n\}$ is a set of representatives for the cosets of $\mathrm{Gal}\,(K/k)/\mathrm{Gal}\,(K/M)$. □

Theorem 2.7. *Let K/k be a normal extension and $K=\bigcup_{i\in I}K_i$, where the K_i are finite extensions of k. Then $\{\mathrm{Gal}\,(K/K_i)\,|\,i\in I\}$ is a system of open and closed neighborhoods of $1\in\mathrm{Gal}\,(K/k)$.*

Proof. Let $i \in I$, and assume that N/k is a finite normal subextension of K/k containing K_i. By Theorem 2.6, $\mathrm{Gal}(K/N)$ has finite index in $\mathrm{Gal}(K/K_i)$. Thus if $\mathrm{Gal}(K/N)$ is open and closed, then so is $\mathrm{Gal}(K/K_i)$. If, on the other hand, N/k is an arbitrary finite normal subextension of K/k, then the existence of primitive elements implies that there is an $i \in I$ such that $N \subseteq K_i$, i.e., $\mathrm{Gal}(K/K_i) \subseteq \mathrm{Gal}(K/N)$. But then $\{\mathrm{Gal}(K/K_i) \mid i \in I\}$ is a neighborhood basis at $1 \in \mathrm{Gal}(K/k)$. □

Theorem 2.8. *Let K/k be a normal extension with subextension M/k. The topology of* $\mathrm{Gal}(K/M)$ *is induced by the topology of* $\mathrm{Gal}(K/k)$*, and* $\mathrm{Gal}(K/M)$ *is closed in* $\mathrm{Gal}(K/k)$*. Moreover,* $\mathrm{Gal}(K/M)$ *is open if and only if M/k is finite.*

Proof. For two subextensions M/k and N/k of K/k, we have

$$\mathrm{Gal}(K/M) \cap \mathrm{Gal}(K/N) = \mathrm{Gal}(K/MN).$$

Therefore, Theorem 2.7 implies that $\{\mathrm{Gal}(K/MN) \mid N/k \text{ finite and normal}\}$ is a neighborhood basis at $1 \in \mathrm{Gal}(K/M)$ for the topology induced by $\mathrm{Gal}(K/k)$, and as well for the topology as the Galois group $\mathrm{Gal}(K/M)$. As a compact group, $\mathrm{Gal}(K/M)$ is closed in $\mathrm{Gal}(K/k)$. If $\mathrm{Gal}(K/M)$ is open, then it has finite index in $\mathrm{Gal}(K/k)$. Theorem 2.6 then easily implies that M/k is finite. □

2.2 The Main Theorem of Galois Theory

Theorem 2.9. *Let K/k be a normal extension. Then $\Phi(M) = \mathrm{Gal}(K/M)$ defines a bijection Φ between the set of all intermediate fields M of K/k and the set of all (closed) subgroups of* $\mathrm{Gal}(K/k)$*.*

The inverse map of Φ is the map Ψ that assigns to each subgroup U of $\mathrm{Gal}(K/k)$ *its fixed field $K(U)$.*

Proof. From finite Galois theory and Theorem 2.5 we deduce that for intermediate fields M, the fixed field of $\mathrm{Gal}(K/M)$ equals M. Thus, it remains to show that $U = \mathrm{Gal}(K/K(U))$ for every subgroup U of $\mathrm{Gal}(K/k)$. Obviously, we have $U \subseteq \mathrm{Gal}(K/K(U))$.

Let $\{K_i \mid i \in I\}$ be the set of all finite normal subextensions of $K/K(U)$. There is a natural commutative diagram

$$\begin{array}{ccc} \mathrm{Gal}(K/K(U)) & \xrightarrow{\ \sim\ } & \varprojlim \mathrm{Gal}(K_i/K(U)) \\ \uparrow & & \uparrow{\scriptstyle \varphi_1} \\ U & \xrightarrow{\ \varphi_2\ } & \varprojlim U|_{K_i} \end{array}$$

Clearly, $K(U)$ is the fixed field of $U|_{K_i}$ in K_i. By finite Galois theory we conclude that $\mathrm{Gal}(K_i/K(U)) = U|_{K_i}$ and that φ_1 is the identity map. Next

φ_2 maps U onto a dense subset of $\varprojlim U|_{K_i}$, hence onto $\varprojlim U|_{K_i}$ because U is closed. Thus φ_2 is an isomorphism, and we find that $U = \mathrm{Gal}(K/K(U))$. □

Let us formulate some theorems that follow immediately from classical Galois theory and the preceding results.

Theorem 2.10. *Let K/k be a normal extension and N/k a normal subextension. Then the natural sequence*

$$1 \longrightarrow \mathrm{Gal}(K/N) \longrightarrow \mathrm{Gal}(K/k) \longrightarrow \mathrm{Gal}(N/k) \longrightarrow 1$$

is an exact sequence of profinite groups.

Theorem 2.11. *Let Ω/k be a normal extension with subextensions K_1/k and K_2/k. If K_1/k is normal, then so is K_1K_2/K_2, and the natural morphism*

$$\mathrm{Gal}(K_1K_2/k) \longrightarrow \mathrm{Gal}(K_1/K_1 \cap K_2)$$

is an isomorphism.

If K_2/k is also normal, then there is a natural isomorphism

$$\mathrm{Gal}(K_1K_2/k) \longrightarrow \mathrm{Gal}(K_1/k) \times_G \mathrm{Gal}(K_2/k),$$

where we have put $G = \mathrm{Gal}(K_1 \cap K_2/k)$.

Example 2.12. If k is a finite field with algebraic closure K, then $\mathrm{Gal}(K/k)$ is isomorphic to the profinite completion of $\mathbb{Z}$ (compare Example 1.7).

3. Cohomology of Profinite Groups

The basis of class field theory and Galois cohomology, which are used for deriving almost all the results on p-extensions in this book, is the calculus of cohomology groups of discrete modules of profinite groups. It is this calculus that we have to build first. Here we mainly have to take the requirements for Galois cohomology into consideration, whereas the cohomology of finite groups that is needed for class field theory is only developed as far as is necessary for the formulation of the theorems that will be important for us. See also J.P. Serre [51, Chap. VI–XI].

3.1 Definition of Cohomology Groups

Let G be a profinite group and A a G-module. We shall define the cohomology groups of G and A using a special complex.

For $n \geq 1$, let $K^n(G, A)$ denote the set of all continuous maps of the n-fold product G^n into A and put $K^0(G, A) = A$. We then transfer the additive structure of A to $K^n(G, A)$.

As for the elements of $M_G(X)$ (see Section 1.5), the fact that $f \in K^n(G, A)$ is continuous means that the function $f(x_1, \ldots, x_n)$ only depends on the cosets of x_ν modulo some open normal subgroup of G.

Now

$$\begin{aligned}(d_n f)(x_1, \ldots, x_{n+1}) = x_1 f(x_2, \ldots, x_{n+1}) & \\ + \sum_{\nu=1}^{n} (-1)^\nu f(x_1, \ldots, x_\nu x_{\nu+1}, \ldots, x_{n+1}) & \\ + (-1)^{n+1} f(x_1, \ldots, x_n) &\end{aligned}$$

defines a homomorphism $d_n : K^n(G, A) \longrightarrow K^{n+1}(G, A)$. We want to show that $K(G, A) = \coprod_{n=0}^{\infty} K^n(G, A)$, together with the map $d = \{d_n \mid n = 0, 1, \ldots\}$, is a complex.

Theorem 3.1. *For $n \geq 1$, we have $d_n d_{n-1} = 0$.*

Proof. We first prove Theorem 3.1 for induced modules $A = M_G(X)$. To this end, we define for each $g \in G$ a homomorphism $s_n(g) : K^n(G, M_G(X)) \longrightarrow K^{n-1}(G, M_G(X))$ by putting

$$((s_n(g)f)(x_1,\dots,x_{n-1},x) = f(g^{-1}x,x_1,\dots,x_{n-1},g) \tag{3.1}$$

for any $f \in K^n(G, M_G(X))$. Here $f : G^{n+1} \longrightarrow X$ is a continuous X-valued function of $n+1$ variables $x_1,\dots,x_n,x$. The first n variables refer to forming the complex, the last one to the induced module. It is immediately checked that

$$d_{n-1}s_n(g)f + s_{n+1}(g)d_nf = f \tag{3.2}$$

for all $n \geq 1$, $g \in G$, and $f \in K^n(G, M_G(X))$.

We now prove $d_nd_{n-1} = 0$ by induction. For $n = 1$, $d_1d_0 = 0$ is checked directly. Now assume that we know $d_md_{m-1} = 0$ for some integer $m \geq 1$. Replace n in (3.2) by $m+1$ and f by d_mf. Then we get

$$s_{m+2}(g)d_{m+1}d_mf = d_mf - d_ms_{m+1}(g)d_mf,$$

and applying (3.2) once more we find

$$s_{m+2}(g)d_{m+1}d_mf = d_md_{m-1}s_m(g)f = 0 \quad \text{for all } g \in G,$$

which, by (3.1), immediately implies $d_{m+1}d_m = 0$.

In order to prove Theorem 3.1 for arbitrary G-modules A, we first remark that

$$(\overline{\psi}f)(h_1,\dots,h_n) = \psi f(\varphi h_1,\dots,\varphi h_n), \quad h_1,\dots,h_n \in H, \tag{3.3}$$

assigns to each morphism $[\varphi,\psi] : [G,A] \longrightarrow [H,B]$ a morphism $\overline{\psi} : K^n(G,A) \longrightarrow K^n(H,B)$ of complexes.

The map

$$[G,A] \longmapsto K(G,A), \quad [\varphi,\psi] \longmapsto \overline{\psi}$$

is a functor from $\mathcal{C}$ to the category of graded groups whose restriction to $\mathcal{C}_G$ is exact.

It follows that the injection $A \hookrightarrow M_G(A)$ (see Example 1.24) induces a commutative diagram

$$\begin{array}{ccccc}
K^{n-1}(G,A) & \xrightarrow{d_{n-1}} & K^n(G,A) & \xrightarrow{d_n} & K^{n+1}(G,A) \\
\downarrow & & \downarrow & & \downarrow \\
K^{n-1}(G,M_G(A)) & \xrightarrow{d_{n-1}} & K^n(G,M_G(A)) & \xrightarrow{d_n} & K^{n+1}(G,M_G(A)),
\end{array}$$

where the vertical arrows denote injections. This immediately implies Theorem 3.1. □

The preceding considerations also imply

Theorem 3.2. *The map*

$$[G,A] \longmapsto K(G,A),$$

together with (3.3), defines a functor from $\mathcal{C}$ to the category of complexes whose restriction to $\mathcal{C}_G$ is exact.

Definition 3.3. We now define the *cohomology groups* $\mathrm{H}^n(G, A)$ of G and A as the cohomology groups of the complex $K^n(G, A)$:

$$\mathrm{H}^n(G, A) = \mathrm{H}^n(K(G, A)) = \begin{cases} \ker d_n / \operatorname{im} d_{n-1}, & \text{if } n \geq 1, \\ \ker d_n, & \text{if } n = 0. \end{cases}$$

For each $n \geq 0$, the morphism $\overline{\psi}$ of complexes induces a homomorphism $\psi^* : \mathrm{H}^n(G, A) \longrightarrow \mathrm{H}^n(H, B)$.

From Theorem 3.2 we deduce, using the cohomology theory of complexes

Theorem 3.4. *The map*

$$[G, A] \longmapsto \mathrm{H}^n(G, A), \quad [\varphi, \psi] \longmapsto \psi^*$$

is a functor from $\mathcal{C}$ to the category of abelian groups.

Theorem 3.5. *Let G and H be profinite groups, and assume that*

$$\begin{array}{ccccccccc} 0 & \longrightarrow & A & \longrightarrow & B & \longrightarrow & C & \longrightarrow & 0 \\ & & \downarrow & & \downarrow & & \downarrow & & \\ 0 & \longrightarrow & A' & \longrightarrow & B' & \longrightarrow & C' & \longrightarrow & 0 \end{array}$$

is a commutative diagram over $\mathcal{C}$, where the top row is an exact sequence of G-modules and the bottom row is an exact sequence of H-modules. Then the diagram

$$\begin{array}{ccccccc} 0 \to \mathrm{H}^0(G, A) & \longrightarrow & \mathrm{H}^0(G, B) & \longrightarrow & \mathrm{H}^0(G, C) & \xrightarrow{\Delta_0} & \mathrm{H}^1(G, A) \to \ldots \\ \downarrow & & \downarrow & & \downarrow & & \downarrow \\ 0 \to \mathrm{H}^0(H, A') & \longrightarrow & \mathrm{H}^0(H, B') & \longrightarrow & \mathrm{H}^0(H, C') & \xrightarrow{\Delta_0} & \mathrm{H}^1(H, A') \to \ldots \end{array}$$

$$\begin{array}{ccccccc} \to \mathrm{H}^{n-1}(G, C) & \xrightarrow{\Delta_{n-1}} & \mathrm{H}^n(G, A) & \longrightarrow & \mathrm{H}^n(G, B) & \longrightarrow & \mathrm{H}^n(G, C) \ldots \\ \downarrow & & \downarrow & & \downarrow & & \downarrow \\ \to \mathrm{H}^{n-1}(H, C') & \xrightarrow{\Delta_{n-1}} & \mathrm{H}^n(H, A') & \longrightarrow & \mathrm{H}^n(H, B') & \longrightarrow & \mathrm{H}^n(H, C') \ldots \end{array}$$

where Δ_{n-1} denotes the connecting homomorphism, *is exact and commutes.*

Theorem 3.6. *Let the G-module A be a direct sum of G-modules A_i, $i \in I$. Then $\mathrm{H}^n(G, A)$ is direct sum of the abelian groups $\mathrm{H}^n(G, A_i)$, $i \in I$.*

The proof of Theorem 3.1 also implies

Theorem 3.7. *Let G be a profinite group and X an abelian group. Then*

$$\mathrm{H}^n(G, M_G(X)) = 0 \quad \textit{for } n \geq 1.$$

Proof. Suppose that $f \in K^n(G, M_G(X))$ and $d_n f = 0$. Then (3.2) implies

$$f = d_{n-1}s_n(g)f.$$

□

Let A denote an arbitrary G-module.

For $n = 0$, we have

$$\mathrm{H}^0(G, A) = A^G = \{a \in A \mid ga = a \text{ for } g \in G\}.$$

This fact, together with Theorems 3.5 and 3.7 characterizes the cohomology groups of G up to isomorphism. This is a consequence of the method of dimension shifting, which we will discuss in Section 3.3.

Here, let us look at the case $n = 1$.

A continuous map $f : G \longrightarrow A$ is called a *crossed homomorphism* if

$$f(g_1 g_2) = f(g_1) + g_1 f(g_2)$$

for all $g_1, g_2 \in G$.

A crossed homomorphism is said to *split* if there exists an $a \in A$ such that

$$f(g) = ga - a.$$

The group $\mathrm{H}^1(G, A)$ is the quotient of the group of all crossed homomorphisms by the group of all split crossed homomorphism. If G acts trivially on A, then $\mathrm{H}^1(G, A)$ is simply the group of continuous homomorphisms.

3.2 Group Extensions

Let A be a finite abelian group and

$$0 \longrightarrow A \longrightarrow \widetilde{G} \longrightarrow G \longrightarrow 1$$

an exact sequence of profinite groups. The theory of group extensions for finite groups can be transferred without problems to the case of profinite groups as long as one observes that all occurring operations are continuous. In particular, the elements of $\mathrm{H}^2(G, A)$ correspond biuniquely to the equivalence classes of group extensions of G by A.

3.3 Dimension Shifting

According to Example 1.24, every G-module A can be embedded into an induced module $M_G(X)$, i.e., there is an exact sequence

$$0 \longrightarrow A \longrightarrow M_G(X) \longrightarrow C \longrightarrow 0$$

of G-modules. According to Theorems 3.5 and 3.7, this implies the exact sequence

$$\mathrm{H}^n(G, M_G(X)) \longrightarrow \mathrm{H}^n(G,C) \longrightarrow \mathrm{H}^{n+1}(G,A) \longrightarrow 0,$$

and for $n \geq 1$

$$\mathrm{H}^n(G,C) \simeq \mathrm{H}^{n+1}(G,A).$$

This allows us to reduce the n-th cohomology groups to the particularly simple cohomology group of dimension 0. This method – called *dimension shifting* – makes it possible to prove many theorems of cohomology theory quite easily.

In general, we have the following principle:

Theorem 3.8. *Let G and H be profinite groups, and let F be an exact covariant functor from the category C_H into the category C_G that maps induced modules to induced modules. Moreover, let λ_m be a functorial morphism defined for all $A \in C_H$, where $\lambda_m(A)$ is a homomorphism $\mathrm{H}^m(G,FA) \longrightarrow \mathrm{H}^m(H,A)$.*

Then there exists a unique family $\{\lambda_n \mid n = m, m+1, \ldots\}$ of functorial morphisms with the property that for all exact sequences

$$0 \longrightarrow A \longrightarrow B \longrightarrow C \longrightarrow 0$$

over C_H and all $n \geq m$ the diagram

$$\begin{array}{ccc} \mathrm{H}^n(G,FC) & \xrightarrow{\lambda_n(C)} & \mathrm{H}^n(H,C) \\ {\scriptstyle \Delta_n}\big\downarrow & & \big\downarrow{\scriptstyle \Delta_n} \\ \mathrm{H}^{n+1}(G,FA) & \xrightarrow{\lambda_{n+1}(A)} & \mathrm{H}^{n+1}(H,A) \end{array} \tag{3.4}$$

commutes.

If λ_m is an isomorphism, then so is λ_n for any $n \geq m$.

Proof. We start by proving the uniqueness of λ_n by induction over n. Let

$$0 \longrightarrow A \xrightarrow{\varphi} M_H(X) \longrightarrow C \longrightarrow 0$$

be an exact sequence of H-modules. By the assumptions of Theorem 3.8, we have an exact and commuting diagram

$$
\begin{array}{ccc}
\mathrm{H}^n(G, FM_H(X)) & \xrightarrow{\lambda_n(M_H(X))} & \mathrm{H}^n(H, M_H(X)) \\
\downarrow & & \downarrow \\
\mathrm{H}^n(G, FC) & \xrightarrow{\lambda_n(C)} & \mathrm{H}^n(H, C) \\
{\scriptstyle \Delta_n}\downarrow & & {\scriptstyle \Delta_n}\downarrow \\
\mathrm{H}^{n+1}(G, FA) & \xrightarrow{\lambda_{n+1}(A)} & \mathrm{H}^{n+1}(H, A) \\
\downarrow & & \downarrow \\
0 & & 0
\end{array}
\tag{3.5}
$$

This implies that λ_{n+1} is uniquely determined for all $n \geq m$ by λ_n. If λ_n is an isomorphism, then so is λ_{n+1}, and this proves the last claim of Theorem 3.8.

The existence of λ_n is also proved by induction. If λ_n is defined for some $n \geq m$, then we define λ_{n+1} with the help of diagram (3.5).

This definition does not depend on the choice of the induced module $M_H(X)$. In fact, assume that

$$
0 \longrightarrow A \xrightarrow{\varphi'} M_H(X') \longrightarrow C' \longrightarrow 0
$$

is another exact sequence of H-modules; then we embed A into

$$
M_H(X \oplus X') \simeq M_H(X) \oplus M_H(X')
$$

using the map $\varphi'' = \varphi \oplus \varphi''$, and we get the exact and commutative diagram

$$
\begin{array}{ccccccccc}
0 & \longrightarrow & A & \longrightarrow & M_H(X) & \longrightarrow & C & \longrightarrow & 0 \\
 & & \uparrow{\scriptstyle \mathrm{id}} & & \uparrow & & \uparrow & & \\
0 & \longrightarrow & A & \longrightarrow & M_H(X \oplus X') & \longrightarrow & C'' & \longrightarrow & 0 \\
 & & \downarrow{\scriptstyle \mathrm{id}} & & \downarrow & & \downarrow & & \\
0 & \longrightarrow & A & \longrightarrow & M_H(X') & \longrightarrow & C' & \longrightarrow & 0
\end{array}
$$

The commutativity of the diagram

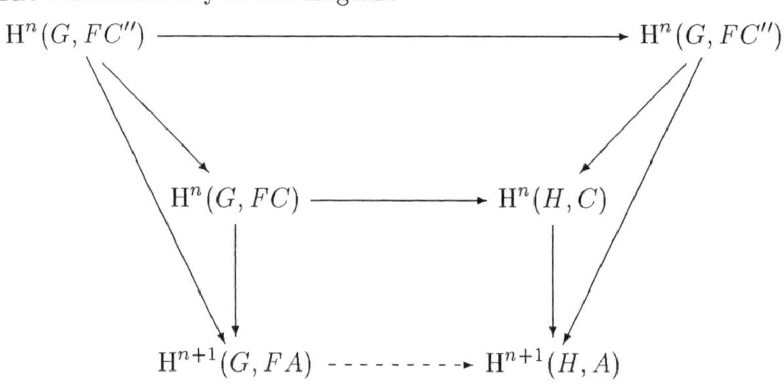

and of the corresponding diagram for C' then implies our claim.

Now let

$$0 \longrightarrow A \longrightarrow B \longrightarrow C \longrightarrow 0$$

be an exact sequence of H-modules. We embed B into $M_H(X)$ and get an exact and commutative diagram

$$\begin{array}{ccccccccc}
0 & \longrightarrow & A & \longrightarrow & B & \longrightarrow & C & \longrightarrow & 0 \\
 & & \big\downarrow{\scriptstyle \mathrm{id}} & & \big\downarrow & & \big\downarrow & & \\
0 & \longrightarrow & A & \longrightarrow & M_H(X) & \longrightarrow & C' & \longrightarrow & 0
\end{array}$$

From the corresponding diagram

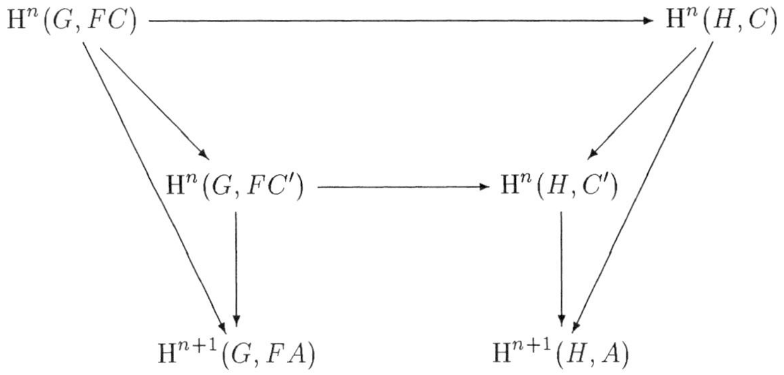
Hⁿ(G, FC)
Hⁿ(H, C)
Hⁿ(G, FC′)
Hⁿ(H, C′)
Hⁿ⁺¹(G, FA)
Hⁿ⁺¹(H, A)

we can read off (3.4).

Starting with the diagram

$$\begin{array}{ccccccccc}
0 & \longrightarrow & A & \longrightarrow & M_H(A) & \longrightarrow & C & \longrightarrow & 0 \\
 & & \big\downarrow & & \big\downarrow & & \big\downarrow & & \\
0 & \longrightarrow & B & \longrightarrow & M_H(B) & \longrightarrow & C' & \longrightarrow & 0
\end{array}$$

we can similarly prove that λ_n is functorial, i.e., that the diagram

$$\begin{array}{ccc}
\mathrm{H}^n(G, FA) & \xrightarrow{\lambda_n(A)} & \mathrm{H}^n(H, A) \\
\big\downarrow & & \big\downarrow \\
\mathrm{H}^n(G, FB) & \xrightarrow{\lambda_n(B)} & \mathrm{H}^n(H, B)
\end{array}$$

commutes. □

We now give some applications of Theorem 3.8.

3.4 Shapiro's Lemma

Let G be a profinite group, H a subgroup of G, and A an H-module. In Example 1.27 we have constructed a morphism $[G, M_G^H(A)] \longrightarrow [H, A]$. This morphism induces, for every $n \geq 0$, a homomorphism $\varphi_n^* : \mathrm{H}^n(G, M_G^H(A)) \longrightarrow \mathrm{H}^n(H, A)$.

Theorem 3.9. *The homomorphisms φ_n^* are isomorphisms.*

Proof. The functor $M_G^H(A) : \mathcal{C}_H \longrightarrow \mathcal{C}_G$ is exact, and $M_G^H(M_H(X))$ is isomorphic to $M_G(X)$. Thus we can apply Theorem 3.8. Since it is easy to check that φ_0^* is an isomorphism, the claim follows. □

Theorem 3.9 allows to compute the cohomology of a subgroup from the cohomology of the group itself.

3.5 Restriction and Corestriction

Let G be a profinite group, H a subgroup of G, and A a G-module. The restriction homomorphism defined in Example 1.26 induces a homomorphism res $: \mathrm{H}^n(G, A) \longrightarrow \mathrm{H}^n(H, A)$ that is also called the *restriction*.

Now assume that the index $(G : H)$ is finite. We want to use Theorem 3.8 (with the roles of G and H switched) to define a map $\mathrm{H}^n(H, A) \longrightarrow \mathrm{H}^n(G, A)$ in the other direction.

As our functor F we take the map that assigns to every G-module A the corresponding H-module. Then F maps $M_G(X)$ to $M_H(M_{G/H}(X))$, where $M_{G/H}(X)$ is the group of all maps $G/H \longrightarrow X$. We define a map $\lambda_0(A) : A^H \longrightarrow A^G$ by

$$\lambda_0(A)a = \sum_{r \in G/H} ra \quad \text{for all } a \in A^H.$$

Clearly λ_0 is a functorial morphism and hence can be extended to higher dimensions. The map cor $: \mathrm{H}^n(H, A) \longrightarrow \mathrm{H}^n(G, A)$ defined in this way is called the *corestriction*.

Let m be an integer and X an abelian group. The endomorphism of X defined by $x \longmapsto mx$ will also be denoted by m. Then

Theorem 3.10. *We have*

$$\mathrm{cor} \circ \mathrm{res} = (G : H). \tag{3.6}$$

Proof. We apply Theorem 3.8 (with the roles of G and H switched) and take F to be the identity map of $\mathcal{C}_G$. It can be checked immediately that that (3.6) holds for the cohomology groups H^0 of dimension 0. The uniqueness assertion of Theorem 3.8 now implies that (3.6) holds for all dimensions. □

Theorem 3.10 implies

Theorem 3.11. *If G is a finite group and A an arbitrary G-module, then for all $n \geq 1$, the cohomology groups $\mathrm{H}^n(G, A)$ are killed by the order of G.*

Proof. We apply Theorem 3.10 in the case $H = \{1\}$. Then for all $\alpha \in \mathrm{H}^n(G, A)$, the restriction $\operatorname{res} \alpha = 0$, hence

$$(G : \{1\})\alpha = \operatorname{cor} \circ \operatorname{res} \alpha = 0.$$

□

3.6 The Transfer

Let H be a subgroup of a finite group G. In algebraic number theory, the corestriction

$$\mathrm{H}^1(H, \mathbb{Q}/\mathbb{Z}) \longrightarrow \mathrm{H}^1(G, \mathbb{Q}/\mathbb{Z}), \tag{3.7}$$

where G acts trivially on $\mathbb{Q}/\mathbb{Z}$, is particularly important. By duality theory for finite abelian groups, there is a map $\mathrm{Ver} : G/[G, G] \longrightarrow H/[H, H]$ corresponding to (3.7) and called the *transfer* (Verlagerung in German) from G to H.

Theorem 3.12. *The transfer from G to H is given explicitly by*

$$\overline{g} \longmapsto \prod_{r \in R} h(g, r),$$

where r runs through a system R of left representatives of G/H and where $h(g, r) \in H$ is defined by $gr = \widetilde{gr}h(g, r)$; here $\widetilde{gr}$ denotes the representative of gr in R.

Proof. From the exact sequence

$$0 \longrightarrow \mathbb{Q}/\mathbb{Z} \longrightarrow M_G(\mathbb{Q}/\mathbb{Z}) \longrightarrow C \longrightarrow 0$$

we get the exact and commutative diagram

$$\begin{array}{ccccc}
\mathrm{H}^0(H, C) & \xrightarrow{\Delta_0} & \mathrm{H}^1(H, \mathbb{Q}/\mathbb{Z}) & \longrightarrow & 0 \\
\downarrow & & \downarrow & & \\
\mathrm{H}^0(G, C) & \xrightarrow[\Delta_0]{} & \mathrm{H}^1(G, \mathbb{Q}/\mathbb{Z}) & \longrightarrow & 0.
\end{array}$$

Let $\chi \in \mathrm{H}^1(H, \mathbb{Q}/\mathbb{Z})$ and let $\overline{f}$ be a preimage of χ under Δ_0 such that $f \in M_G(\mathbb{Q}/\mathbb{Z})$. Then

$$f(xh) - f(x) = \chi(h)$$

for all $x \in G$. We have cor $\overline{f} = \sum_{r \in R} r\overline{f}$, and the corresponding character cor $\chi = \chi'$ is given by

$$\chi'(g) = \sum_{r \in R} rf(g) - rf(1) = \sum_{r \in R} f(gr) - f(r).$$

Since

$$\sum_{r \in R} f(r) = \sum_{r \in R} f(\widetilde{gr}),$$

we have

$$\sum_{r \in R} f(gr) - f(r) = \sum_{r \in R} f(gr) - f(\widetilde{gr}) = \sum_{r \in R} \chi(h(g,r)).$$

Thus

$$\chi'(g) = \chi\Big(\prod_{r \in R} h(g,r)\Big).$$

This implies the claim. □

The functorial properties of cor now imply analogous properties of the transfer.

3.7 Inflation and Transgression

Let G be a profinite group, H a normal subgroup of G, and A a G-module. The morphism $[G/H, A^H] \longrightarrow [G, A]$ considered in Example 1.26 induces a homomorphism inf between the corresponding cohomology groups that will be called *inflation*.

Moreover, we can define an automorphism of $\mathrm{H}^n(H, A)$ for every $g \in G$: the morphisms $\psi : a \longmapsto ga$ for $a \in A$ and $\varphi : h \to g^{-1}hg$ for $h \in H$ are compatible and induce an endomorphism $\widetilde{g}^H = \widetilde{g}$ of $\mathrm{H}^n(H, A)$. For all $g_1, g_2 \in G$ we have

$$\widetilde{g_1} \cdot \widetilde{g_2} = \widetilde{g_1 g_2},$$

and this implies that $\widetilde{g}$ is an endomorphism of $\mathrm{H}^n(H, A)$.

If $g \in H$, or if g acts trivially on A, then $\widetilde{g}$ is the identity map in dimension 0. By Theorem 3.8, the same must be true for all dimensions. This implies that we can regard $\mathrm{H}^n(H, A)$ as a discrete G/H-module, and we have the following

Theorem 3.13. *The elements in the image of* res $: \mathrm{H}^n(G, A) \longrightarrow \mathrm{H}^n(H, A)$ *are invariant under the action of* G/H.

Proof. For every $g \in G$, we have a commutative diagram

$$\begin{array}{ccc} \mathrm{H}^n(G,A) & \xrightarrow{\mathrm{res}} & \mathrm{H}^n(H,A) \\ \widetilde{g}^G \Big\downarrow & & \widetilde{g}^H \Big\downarrow \\ \mathrm{H}^n(G,A) & \xrightarrow{\mathrm{res}} & \mathrm{H}^n(H,A). \end{array}$$

$\widetilde{g}^G$ is the identity map. This implies the claim. □

Now we define a functorial morphism

$$\mathrm{tra} : \mathrm{H}^n(H,A)^{G/H} \longrightarrow \mathrm{H}^{n+1}(G/H, A^H)$$

(called *transgression*) for $n \geq 1$ under the assumption that

$$\mathrm{H}^\nu(H,A) = 0 \quad \text{for } \nu = 1, \dots, n-1.$$

Consider the case $n = 1$ first and let $\overline{a} \in \mathrm{H}^1(H,A)^{G/H}$, where a is a crossed homomorphism. We want to construct a lift b of a to G with the following properties:

(i) $b : G \longrightarrow A$ is continuous;

(ii) $gb(g^{-1}hg) - b(h) = hb(g) - b(g)$ for all $g \in G, h \in H$;

(iii) $b(hg) = b(h) + hb(g)$ for all $g \in G, h \in H$.

There exists an open normal subgroup U of G such that $a(h)$ only depends on the cosets of $H/H \cap U$ and such that, for all $h \in H$, $a(h)$ is invariant under U. Moreover, according to Theorem 1.16 there is a continuous section $s : G/H \longrightarrow G$. Since $\overline{a}$ is invariant under G, for $\gamma \in G/H$ there exist elements $b(s\gamma) \in G$ such that

$$s\gamma a((s\gamma)^{-1}hs\gamma) - a(h) = hb(s\gamma) - b(s\gamma) \quad \text{for all } h \in H. \tag{3.8}$$

The left hand side of (3.8) is constant for all elements $s\gamma$ from a fixed coset mod U because of the choice of U. Thus, for $b(s\gamma)$ we can chose one and the same element for all $s\gamma$ in the same coset G/U, which secures the continuity of b as a function of γ. For arbitrary $g = hs\gamma \in G$ we put

$$b(g) = a(h) + hb(s\gamma).$$

Then $b(g)$ is continuous by construction and satisfies ii) and iii), as is easily checked.

We now put

$$f(g_1, g_2) = b(g_1) + g_1 b(g_2) - b(g_1 g_2). \tag{3.9}$$

Then $f(g_1, g_2)$ is invariant under $h \in H$. In fact, ii) implies

$$
\begin{aligned}
hf(g_1,g_2) - f(g_1,g_2) = g_1 b(g_1^{-1}hg_1) - b(h) \\
+ g_1[g_2 b(g_2^{-1}g_1^{-1}hg_1g_2) - b(g_1^{-1}hg_1)] \\
- [g_1g_2 b((g_1g_2)^{-1}hg_1g_2) - b(h)] = 0
\end{aligned}
$$

Using the invariance of $f(g_1, g_2)$ under the action of $h \in H$ we can also check that, for all $h_1, h_2 \in H$ and $g_1, g_2 \in G$, we have

$$f(h_1g_1, h_2g_2) = f(g_1, g_2). \tag{3.10}$$

By the way $f(g_1, g_2)$ is built, it is clear that

$$\varphi(g_1, g_2) = f(s\gamma_1, s\gamma_2) \tag{3.11}$$

defines a factor set φ with $\overline{\varphi} \in \mathrm{H}^2(G/H, A^H)$.

The class $\overline{\varphi}$ does not depend on the choice of s or $b(s\gamma)$. In fact, for

$$s'\gamma = h\gamma s\gamma,$$

$b'(s'\gamma)$ has the form

$$b'(s'\gamma) = b(s'\gamma) + c(\gamma), \quad c(\gamma) \in A^H.$$

This implies

$$b'(g) = b(g) + c(\overline{g})$$

for all $g \in G$ ($\overline{g}$ denotes the class of g in G/H). Then

$$
\begin{aligned}
\varphi'(\gamma_1, \gamma_2) &= f'(s'\gamma_1, s'\gamma_2) \\
&= b(s'\gamma_1) + s'\gamma_1 b(s'\gamma_2) - b(s'\gamma_1 s'\gamma_2) + c(\gamma_1) + s'\gamma_1 b(\gamma_2) - b(\gamma_1\gamma_2).
\end{aligned}
$$

Thus φ and φ' differ by a split factor system, which is what we wanted to show.

Now we define the transgression tra in dimension 1 by

$$\mathrm{tra}\, \overline{a} = \overline{\varphi}. \tag{3.12}$$

Clearly tra is a homomorphism $\mathrm{H}^1(H, A)^{G/H} \longrightarrow \mathrm{H}^2(G/H, A^H)$. We now show that the sequence

$$
\begin{array}{ccccccc}
0 \longrightarrow & \mathrm{H}^1(G/H, A^H) & \xrightarrow{\mathrm{inf}} & \mathrm{H}^1(G, A) & \xrightarrow{\mathrm{res}} & \mathrm{H}^1(H, A)^{G/H} \\
& & & & & \downarrow \mathrm{tra} \\
& & & \mathrm{H}^2(G, A) & \xleftarrow{\mathrm{inf}} & \mathrm{H}^2(G/H, A^H)
\end{array}
$$

is exact.

Exactness at $\mathrm{H}^1(G/H, A^H)$: if $\overline{a} \in \ker \mathrm{inf}$, then there is a $c \in A$ such that $a(\overline{g}) = gc - c$ for $g \in G$. Since $a(\overline{g}) = a(\overline{gh}) = ghc - c$ for all $h \in H$, we have $c \in A^H$, hence a splits.

Exactness at $\mathrm{H}^1(G,A)$: if $\overline{a} \in \ker \mathrm{res}$, then there is a $c \in A$ such that $a(h) = hc - c$ for all $h \in H$. Then $a'(g) = a(g) - gc + c$ defines a crossed homomorphism homologous to a which vanishes on H. Therefore, a' only depends on the cosets of G/H and is invariant under H, hence $\overline{a} = \overline{a'}$ is in the image of inf.

Exactness at $\mathrm{H}^1(H,A)^{G/H}$: if $\overline{a} \in \mathrm{im}\ \mathrm{res}$, i.e., if a is the restriction of a crossed homomorphism b to H. then b satisfies the conditions i) to iii), and the factor system assigned to b by (3.11) vanishes; this implies $\mathrm{tra}\ \overline{a} = 0$.

If $a \in \ker \mathrm{tra}$, on the other hand, then (3.9) splits:

$$b(g_1) + g_1 b(g_2) - b(g_1 g_2) = c(\overline{g}_1) + g_1 c(\overline{g}_2) - c(\overline{g}_1 \overline{g}_2).$$

Thus $b' = b - c$ is a crossed homomorphism on G with $\mathrm{res}\ \overline{b}' = \overline{a}$.

Exactness at $\mathrm{H}^2(G/H, A^H)$: if $\overline{\varphi} \in \mathrm{im}\ \mathrm{tra}$, that is, if φ is of the form (3.11), then we clearly have $\inf \overline{\varphi} = 0$.

If $\overline{\varphi} \in \ker \inf$, then

$$\varphi(\overline{g}_1 \overline{g}_2) = c(g_1) + g_1 c(g_2) - c(g_1 g_2).$$

Without loss of generality we may assume $\varphi(1,\overline{g}) = \varphi(\overline{g},1) = 0$ for all $\overline{g} \in G/H$. As is easily checked, the elements $c(g)$ satisfy the conditions i) to iii), and the restriction of c to H is a crossed homomorphism invariant under G/H, that is, $\mathrm{tra}\ \overline{a} = \overline{\varphi}$.

Now assume that

$$\mathrm{H}^i(H,A) = 0 \quad \text{for } 1 \le i \le n-1 \tag{3.13}$$

and that the sequence

$$0 \longrightarrow A \longrightarrow M_G(A) \longrightarrow C \longrightarrow 0$$

is exact. The module $M_G(A)$ is also H-induced. Since $\mathrm{H}^1(H,A) = 0$ for $n \ge 2$, the sequence

$$0 \longrightarrow A^H \longrightarrow M_G(A)^H \longrightarrow C^H \longrightarrow 0$$

is exact, and $M_G(A)^H$ is G/H-induced.

We now define the transgression for dimension n inductively as the map tra that makes the diagram

$$\begin{array}{ccc}
\mathrm{H}^{n-1}(H,C) & \xrightarrow{\mathrm{tra}} & \mathrm{H}^n(G/H,C^H) \\
{\scriptstyle \Delta_{n-1}}\downarrow & & \downarrow{\scriptstyle \Delta_n} \\
\mathrm{H}^n(H,A) & \xrightarrow{\mathrm{tra}} & \mathrm{H}^{n+1}(G/H,A^H)
\end{array}$$

commutative; here the connecting homomorphisms Δ_{n-1} and Δ_n are isomorphisms. By (3.13), C satisfies the induction assumptions.

The theorem of Hochschild-Serre 3.14 now follows from the commutativity of the following diagram:

$$\begin{array}{ccc}
 & & 0 \\
 & & \downarrow \\
\mathrm{H}^{n-1}(G/H, C^H) & \longrightarrow & \mathrm{H}^n(G/H, A^H) \\
\downarrow \scriptstyle{\inf} & & \downarrow \scriptstyle{\inf} \\
\mathrm{H}^{n-1}(G, C) & \longrightarrow & \mathrm{H}^n(G, A) \\
\downarrow \scriptstyle{\mathrm{res}} & & \downarrow \scriptstyle{\mathrm{res}} \\
\mathrm{H}^{n-1}(H, C)^{G/H} & \longrightarrow & \mathrm{H}^n(H, A)^{G/H} \\
\downarrow \scriptstyle{\mathrm{tra}} & & \downarrow \scriptstyle{\mathrm{tra}} \\
\mathrm{H}^n(G/H, C^H) & \longrightarrow & \mathrm{H}^n(G/H, A^H) \\
\downarrow \scriptstyle{\inf} & & \downarrow \scriptstyle{\inf} \\
\mathrm{H}^n(G, C) & \longrightarrow & \mathrm{H}^{n+1}(G, A)
\end{array}$$

Theorem 3.14. *If* $\mathrm{H}^i(H, A) = 0$ *for* $1 \le i \le n-1$, *then the sequence*

$$\begin{array}{ccccccc}
0 & \longrightarrow & \mathrm{H}^n(G/H, A^H) & \xrightarrow{\inf} & \mathrm{H}^n(G, A) & \xrightarrow{\mathrm{res}} & \mathrm{H}^n(H, A)^{G/H} \\
 & & & & & & \downarrow \scriptstyle{\mathrm{tra}} \\
 & & & & \mathrm{H}^{n+1}(G, A) & \xleftarrow{\inf} & \mathrm{H}^{n+1}(G/H, A^H)
\end{array}$$

is exact.

It can be easily seen that the transgression is a functorial morphism. We now deduce some consequences from Theorem 3.14:

Theorem 3.15. *Let* G *be a profinite group,* H *a normal subgroup of finite index, and* A *a* G*-module whose elements have finite order coprime to* $(G : H)$. *Then*

$$\mathrm{H}^n(G/H, A^H) = 0$$

for all $n \ge 1$, *and the restriction*

$$\mathrm{res} : \mathrm{H}^n(G, A) \longrightarrow \mathrm{H}^n(H, A)^{G/H}$$

is an isomorphism.

Proof. By Theorem 3.11, $\mathrm{H}^n(G/H, A^H)$ is killed by $(G : H)$. By assumption, every element of A and therefore every element of $\mathrm{H}^n(G/H, A^H)$ has order coprime to $(G : H)$. This implies the first claim.

In dimension $n = 1$, the second claim follows immediately from Theorem 3.14; for $n > 1$, we prove it by dimension shifting.

If the sequence

$$0 \longrightarrow A \longrightarrow M_G(A) \longrightarrow C \longrightarrow 0$$

is exact, then for every $n \geq 1$ we have commutative diagrams

$$\begin{array}{ccc} \mathrm{H}^n(G,C) & \longrightarrow & \mathrm{H}^n(H,C)^{G/H} \\ {\scriptstyle \Delta_n}\downarrow & & \downarrow{\scriptstyle \Delta_n} \\ \mathrm{H}^{n+1}(G,A) & \longrightarrow & \mathrm{H}^{n+1}(H,A)^{G/H}, \end{array}$$

where the connecting homomorphisms are isomorphisms. Since C also satisfies the conditions of Theorem 3.15, the claim follows by induction. □

3.8 Inductive Limits of Cohomology Groups

As another application of dimension shifting we prove a theorem that is of fundamental importance for the computation of cohomology groups of profinite groups.

Theorem 3.16. *If $\{I, [G_i, A_i], [\varphi_i^j, \psi_i^j]\}$ is an inductive system in the category $\mathcal{C}$ with limit $[G, A]$, then $\{I, \mathrm{H}^n(G_i, A_i), \psi_i^{j\,*}\}$ is an inductive system of abelian groups for every $n \geq 0$, and the natural map*

$$\varinjlim \mathrm{H}^n(G_i, A_i) \longrightarrow \mathrm{H}^n(G, A)$$

is an isomorphism.

Proof. We prove Theorem 3.16 by induction. Assume first that $n = 0$. The map

$$\varinjlim A_I^{G_i} \longrightarrow A^G$$

is obviously injective.

On the other hand, let $\overline{a}_j \in A^G$ and let U_j be the set of all elements of G_j that leave a_j fixed. Then U_j is a subgroup of G_j with finite index. Let $g_j^{(1)}, \ldots, g_j^{(s)}$ be a system of representatives for the left cosets of G_j/U_j. Since $\overline{a}_j$ is invariant under G, there is a $k \geq j$ such that

$$\psi_i^k g_j^{(\nu)} a_j = \psi_j^k a_j \quad \text{for } \nu = 1, \ldots, s.$$

Now $\psi_j^k a_j$ is invariant under G. For if $g_k \in G$ and $\varphi_j^k g_k = g_j^{(\nu)} u$, $u \in U_j$, then

$$g_k \psi_j^k a_j = \psi_i^k g_j^{(\nu)} u a_j = \psi_j^k a_j.$$

Therefore $\overline{a}_j = \overline{\psi_j^k a_j} \in \varinjlim A_i^{G_i}$.

Now assume that Theorem 3.16 is proved for some $n \geq 0$. From the commutative diagram

$$\begin{array}{ccccccccc}
0 & \longrightarrow & A_i & \longrightarrow & M_{G_i}(A_i) & \longrightarrow & C_i & \longrightarrow & 0 \\
 & & \Big\downarrow{\scriptstyle \varphi_i^j} & & \Big\downarrow & & \Big\downarrow & & \\
0 & \longrightarrow & A_j & \longrightarrow & M_{G_j}(A_j) & \longrightarrow & C_j & \longrightarrow & 0 \\
 & & \Big\downarrow{\scriptstyle \varphi_j} & & \Big\downarrow & & \Big\downarrow & & \\
0 & \longrightarrow & A & \longrightarrow & M_G(A) & \longrightarrow & C & \longrightarrow & 0
\end{array}$$

with exact rows we get the exact and commutative diagram

$$\begin{array}{ccccc}
\varinjlim \mathrm{H}^n(G_i, M_{G_i}(A_i)) & \longrightarrow & \varinjlim \mathrm{H}^n(G_i, C_i) & \longrightarrow & \varinjlim \mathrm{H}^{n+1}(G_i, A_i) \longrightarrow 0 \\
\Big\downarrow & & \Big\downarrow & & \Big\downarrow \\
\mathrm{H}^n(G, M_G(A)) & \longrightarrow & \mathrm{H}^n(G, C) & \longrightarrow & \mathrm{H}^{n+1}(G, A) \longrightarrow 0
\end{array}$$

by observing that $\varinjlim M_{G_i}(A_i) = M_G(A)$. By Theorem 1.34, this implies our claim for $n+1$. □

We now give two examples for the computation of cohomology groups using Theorem 3.16.

Theorem 3.17. *If K/k is a normal extension, then*

$$\mathrm{H}^n(\mathrm{Gal}\,(K/k), K^+) = \{0\}$$

for $n \geq 1$.

Proof. If K/k is a finite extension, then the normal basis theorem implies that K^+ is an induced $\mathrm{Gal}\,(K/k)$-module (cf. the remark at the end of Example 1.25). The claim now follows from Theorem 3.7.

If K/k is an arbitrary normal extension, then the $\mathrm{Gal}\,(K/k)$-module K^+ is the inductive limit of the system $\{[\mathrm{Gal}\,(N/k), N^+], N \in \mathcal{N}\}$ (see Theorem 2.2). The claim now follows from Theorem 3.16. □

Theorem 3.18 (Hilbert's Theorem 90). *If K/k is a normal extension, then*

$$\mathrm{H}^1(\mathrm{Gal}\,(K/k), K^\times) = \{0\}. \tag{3.14}$$

Proof. We put $G = \mathrm{Gal}\,(K/k)$. Assume for the moment that K/k is finite, and let f be a crossed homomorphism such that $\overline{f} \in \mathrm{H}^1(G, K^\times)$. Consider the elements

$$b(a) = \sum_{g \in G} f(g) ga \quad \text{for} \quad a \in K^\times.$$

As is easily checked, this implies that

$$hb(a)f(h) = b(a)$$

for all $h \in G$. For the proof of (3.14) it is therefore sufficient to prove the existence of an $a \in K^\times$ such that $b(a) \neq 0$.

Let $1, a, \ldots, a^{n-1}$ be a basis of K/k. The Vandermonde determinant $|ga^\nu|_{g,\nu}$ of the system of equations

$$\sum_{g \in G} x_g g a^\nu = 0, \quad \nu = 0, \ldots, n-1,$$

is $\neq 0$, hence the claim holds for finite K/k.

For arbitrary normal extensions the claim follows again from Theorem 3.16. □

Theorem 3.19. *Assume that m is a natural number, $G = \varprojlim_{i \in I} G_i$, and that A is an arbitrary G-module. If the orders of the groups G_i are finite and coprime to m, then so are the orders of the elements of $\mathrm{H}^n(G, A)$ for $n \geq 1$.*

Proof. Let U_i denote the kernel of the natural morphism $G \longrightarrow G_i$. Then the G-module A is the inductive limit of the system $\{[G_i, A^{U_i}] \mid i \in I\}$.

Thus Theorem 3.19 follows immediately from Theorem 3.11 and 3.16. □

3.9 Cup Products

Let G be a profinite group, and assume that A, B and C are G-modules. Moreover, let $\circ$ be a bilinear map $A \times B \longrightarrow C$ such that

$$g(a \circ b) = ga \circ gb$$

for all $g \in G$, $a \in A$ and $b \in B$. Such a map is called a *pairing* of A and B into C. We shall also write $A \circ B$ for C.

Example 3.20. Let $A = B$ be a ring on which G acts trivially. Then multiplication in A is a pairing.

Example 3.21. Let A and B be arbitrary G-modules. We can make the tensor product $A \otimes_{\mathbb{Z}} B$ into a G-module by putting

$$g(a \otimes b) = ga \otimes gb \quad \text{for } a \in A, b \in B.$$

Then $\otimes$ is a pairing of A and B into $A \otimes_{\mathbb{Z}} B$.

To each pairing of A and B into C there is a bilinear map $\mathrm{H}^n(G,A) \times \mathrm{H}^m(G,B) \longrightarrow \mathrm{H}^{n+m}(G,C)$ for every $n \geq 0$, $m \geq 0$ which is called *cup product* and which we are going to define now.

For $f \in K^n(G,A)$ and $f' \in K^m(G,B)$ we put

$$(f \circ f')(x_1, \ldots, x_{n+m}) = f(x_1, \ldots, x_n) \circ x_1 \cdots x_n f'(x_{n+1}, \ldots x_{n+m}).$$

Then $f \circ f' \in K^{n+m}(G,C)$, and it is easily verified that

$$d_{n+m}(f \circ f') = d_n f \circ f' + (-1)^n f \circ d_m f'.$$

Therefore,

$$\overline{f} \cup \overline{f'} = \overline{f \circ f'}$$

induces a bilinear map $\mathrm{H}^n(G,A) \times \mathrm{H}^m(G,B) \longrightarrow \mathrm{H}^{n+m}(G,C)$ called the *cup product.* Some properties of the cup product follow immediately from the definition of the cup product.

Theorem 3.22. *Let G be a profinite group and $\circ$ a pairing of the G-modules A and B into C. Then*

$$\alpha \cup \beta = \alpha \circ \beta$$

for $\alpha \in \mathrm{H}^0(G,A)$ and $\beta \in \mathrm{H}^0(G,B)$.

Theorem 3.23. *Let G be a profinite group,*

$$0 \longrightarrow A \longrightarrow A' \longrightarrow A'' \longrightarrow 0$$

an exact sequence of G-modules and B a G-module such that

$$0 \longrightarrow A \otimes_{\mathbb{Z}} B \longrightarrow A' \otimes_{\mathbb{Z}} B \longrightarrow A'' \otimes_{\mathbb{Z}} B \longrightarrow 0$$

is also exact. Then

$$(\Delta_n \alpha) \cup \beta = \Delta_{n+m}(\alpha \cup \beta)$$

for $\alpha \in \mathrm{H}^n(G,A'')$ and $\beta \in \mathrm{H}^m(G,B)$, where Δ_n and Δ_{n+m} denote the connecting homomorphisms.

Theorem 3.24. *Let G be a profinite group,*

$$0 \longrightarrow B \longrightarrow B' \longrightarrow B'' \longrightarrow 0$$

an exact sequence of G-modules, and A a G-module such that

$$0 \longrightarrow A \otimes_{\mathbb{Z}} B \longrightarrow A \otimes_{\mathbb{Z}} B' \longrightarrow A \otimes_{\mathbb{Z}} B'' \longrightarrow 0$$

is also an exact sequence. Then, for $\alpha \in \mathrm{H}^n(G,A)$ and $\beta \in \mathrm{H}^m(G,B'')$, we have

$$\alpha \cup \Delta_m \beta = (-1)^n \Delta_{n+m}(\alpha \cup \beta).$$

The functorial properties of the cup product are given by the following theorem.

Theorem 3.25. *Let G, H be profinite groups, and let $\circ$ and $\cdot$ be pairings of the G-modules A and B into C and of the H-modules A' and B' into C', respectively. Let $[\varphi, \psi_A] : [G, A] \longrightarrow [H, A']$, $[\varphi, \psi_B] : [G, B] \longrightarrow [H, B']$ and $[\varphi, \psi_C] : [G, C] \longrightarrow [H, C']$ be morphisms such that*

$$\psi_C(a \circ b) = \psi_A a \cdot \psi_B b \quad \textit{for all } a \in A, b \in B.$$

Then

$$\psi_C^*(\alpha \cup \beta) = \psi_A^* \alpha \cup \psi_B^* \beta$$

for all $\alpha \in \mathrm{H}^n(G, A)$ and $\beta \in \mathrm{H}^m(G, B)$.

Theorems 3.22 to 3.25 uniquely determine the cup product for the pairing given by the tensor product.

This follows again from the method of dimension shifting, which can be applied because if

$$0 \longrightarrow A \longrightarrow M_G(A) \longrightarrow C \longrightarrow 0$$

is an exact sequence, then so is

$$0 \longrightarrow A \otimes_{\mathbb{Z}} B \longrightarrow M_G(A) \otimes_{\mathbb{Z}} B \longrightarrow C \otimes_{\mathbb{Z}} B \longrightarrow 0.$$

Since the tensor product is a right exact functor, we only have to show that $A \otimes_Z B \longrightarrow M_G(A) \otimes_{\mathbb{Z}} B$ is injective. The map $M_G(A) \longrightarrow A$ considered in Example 1.27 composed with $A \longrightarrow M_G(A)$ is the identity map. This implies the claim.

Theorem 3.26. *Let G be a profinite group, and let A, B and C be G-modules. Assume that the pairings $(A \circ B) \circ C = A \circ (B \circ C)$ are defined, and that*

$$(a \circ b) \circ c = a \circ (b \circ c) \quad \textit{for all } a \in A, b \in B, c \in C.$$

Then

$$(\alpha \cup \beta) \cup \gamma = \alpha \cup (\beta \cup \gamma)$$

for all $\alpha \in \mathrm{H}^n(G, A)$, $\beta \in \mathrm{H}^m(G, B)$ and $\gamma \in \mathrm{H}^l(G, C)$.

Theorem 3.27. *Let G be a profinite group, and let A and B be G-modules. Assume that the pairings $A \circ B = B \circ A$ are defined, and that*

$$a \circ b = b \circ a \quad \textit{for all } a \in A, b \in B.$$

Then

$$\alpha \cup \beta = (-1)^{nm} \beta \cup \alpha$$

for $\alpha \in \mathrm{H}^n(G, A)$ and $\beta \in \mathrm{H}^m(G, B)$.

Because of the basic functorial property of the tensor product it suffices to prove Theorems 3.26 and 3.27 for tensor products with corresponding identifications. In this case, they follow without problems using the method of dimension shifting.

Example 3.28. The assumptions of Theorems 3.26 and 3.27 are obviously satisfied if $A = B = C$ is a commutative ring on which G acts trivially.

4. Free pro-p Groups

In the following, p shall always denote a prime number. We restrict our considerations to those profinite groups that are of interest to us as Galois groups of p-extensions, namely pro-p groups. A pro-p group is a profinite group that can be realized as a profinite limit of finite p-groups. We start with a detailed discussion of free prop-p groups.

4.1 Construction of Free pro-p Groups

Definition 4.1. Let G be a pro-p group. A *system of generators* of G is a subset E of G with the following properties:

(i) G is the smallest (closed) subgroup containing E;

(ii) every neighborhood of $1 \in G$ contains almost all (all except finitely many) elements of E.

We shall see below that systems of generators exist in any pro-p group.

Definition 4.2. A system E of generators is called *minimal* if no proper subset of E is a system of generators.

Now we construct free pro-p groups.

Let I be an index set and F_I the free group on the generators $\{s_i \mid i \in I\}$. Moreover, let $\mathfrak{U}$ be the set of normal subgroups N of F_I with the following properties:

(i) $(F_I : N)$ is a power of p;

(ii) almost all elements of $\{s_i \mid i \in I\}$ are in N.

Then $\{F_I/N \mid N \in \mathfrak{U}\}$ is a projective system in the sense of Example 1.6 whose projective limit is a pro-p group.

The group $F(I)$ is called the *free pro-p group with system of generators* $\{s_i \mid i \in I\}$. We will justify this name later on.

The map

$$\varphi : w \longmapsto \prod_N wN, \quad w \in F_I$$

is a homomorphism $F_I \longrightarrow F(I)$ whose image is dense in $F(I)$. Before we can show that φ is injective, we have to prove some auxiliary results.

4.2 The Magnus Group Algebra

Definition 4.3. Let Λ be a ring with a unit element and I some index set. The *Magnus algebra* $\Lambda(I)$ in the variables x_i, $i \in I$, over Λ is the algebra of formal noncommutative (associative) power series in the x_i, $i \in I$, with coefficients in Λ.

We define a homomorphism ψ from the free group F_I on the generators $\{s_i \mid i \in I\}$ to the unit group of $\Lambda(I)$ by putting

$$\psi(s_i) = 1 + x_i.$$

In particular, we have

$$\psi(s_i^{-1}) = \sum_{\nu=0}^{\infty}(-x_i)^{\nu}.$$

Lemma 4.4. *The map ψ is injective.*

Proof. Let

$$s_{i_1}^{a_1} \cdots s_{i_\kappa}^{a_\kappa} \neq 1$$

be an element of F_I in simplified presentation, that is, $i_\nu \neq i_{\nu+1}$ for $\nu = 1, \ldots, \kappa - 1$. Then we have to show that

$$\psi(s_{i_1}^{a_1} \cdots s_{i_\kappa}^{a_\kappa}) = (1 + x_{i_1})^{a_1} \cdots (1 + x_{i_\kappa})^{a_\kappa} \neq 1.$$

The binomial formula gives

$$(1 + x_{i_1})^{a_1} \cdots (1 + x_{i_\kappa})^{a_\kappa} = \sum x_{i_1}^{b_1} \cdots x_{i_\kappa}^{b_\kappa} \prod_{\nu=1}^{\kappa} \binom{a_\nu}{b_\nu}, \tag{4.1}$$

where the sum is over all κ-tuples $[b_1, \ldots, b_\kappa]$ of integers.

Let n denote the characteristic of Λ, and let q be a prime divisor of n. If we let b_ν be the highest power of q dividing a_ν, then $\binom{a_\nu}{b_\nu} \not\equiv 0 \bmod q$, and at least one term in (4.1) does not vanish. □

Theorem 4.5. *The map φ defined in Section 4.1 is injective.*

Proof. Since $\ker \varphi = \{w \in F_I \mid w \in N \text{ for all } N \in \mathfrak{U}\}$, the claim is equivalent to

$$\bigcap \mathfrak{U} = \{1\}.$$

Assume first that I is finite. Let B denote the ideal in the Magnus algebra $\mathbb{F}_p(I)$ consisting of all power series whose constant terms vanish; more generally, let B^ν be the ideal of all power series all of whose terms have at least degree ν. Clearly $\# \mathbb{F}_p(I)/B^\nu$ is a power of p for all $\nu \geq 0$, and $\bigcap_{\nu=1}^{\infty} B^\nu = \{0\}$.

Now put

$$N_\nu = \{w \in F_I \mid \psi(w) - 1 \in B^\nu\}. \tag{4.2}$$

Then N_ν is a normal subgroup of F_I, and Lemma 4.4 implies that $\bigcap N_\nu = \{1\}$. We now show by induction on ν that the index of N_ν in F_I is a power of p. For $\nu = 1$, we have $N_1 = F_I$. Moreover, ψ induces a monomorphism $N_\nu/N_{\nu+1} \longrightarrow B^\nu/B^{\nu+1}$. This implies the claim.

If I is infinite, then we consider the set of all normal subgroups $N_{\nu,J}$ of F_I generated by the sets

$$N_\nu(J), \quad \{s_i \mid i \in I \setminus J\},$$

where J is a finite subset of I and $N_\nu(J)$ the group (4.2) for the index set J. Clearly, $N_{\nu,J} \in \mathfrak{U}$, and we have

$$\bigcap_{\nu,J} N_{\nu,J} = \{1\}.$$

This implies the claim. □

Thus we have an injection $\varphi : F_I \hookrightarrow F(I)$, and we identify F_I with $\varphi(F_I)$. In particular, $\{s_i \mid i \in I\}$ is a minimal set of generators of $F(I)$.

4.3 Abelian pro-p Groups

Under Pontryagin duality between compact and discrete groups, abelian pro-p groups correspond to discrete abelian p-primary torsion groups.

Let G be an abelian pro-p group with exponent p. Then its dual group is a discrete $\mathbb{F}_p$-vector space V. Moreover, let $\{\chi_i \mid i \in I\}$ be a basis of V. The elements $s_i \in G$, $i \in I$ with

$$\langle s_i, \chi_j \rangle = \frac{1}{p} \cdot \delta_{ij}$$

form a minimal system of generators of G, and G is isomorphic to $\prod_I \mathbb{F}_p$.

We can make $\mathbb{Z}_p$ act on an arbitrary pro-p group G as follows: for $a = \lim_{\nu\to\infty} a_\nu \in \mathbb{Z}_p$, $a_\nu \in \mathbb{Z}$, and for $g \in G$ we define

$$g^a = \lim_{\nu\to\infty} g^{a_\nu}.$$

It is easily checked that this definition does not depend on the choice of the sequence a_ν. Moreover, we have

$$\begin{aligned} g^{a+b} &= g^a g^b, \\ (g^a)^b &= g^{ab} \quad \text{for } g \in G, a, b \in A, \end{aligned}$$

and if G is abelian, we also have

$$(g_1 g_2)^a = g_1^a g_2^a \quad \text{for } g_1, g_2 \in G,\ a \in \mathbb{Z}_p.$$

4.4 First Characterization of Free pro-p Groups

We start by proving a few results on pro-p groups that we shall need below.

Theorem 4.6. *Let $F(I)$ be the free pro-p group with system of generators $\{s_i \mid i \in I\}$, let G be a pro-p group, and $\{t_i \mid i \in I\}$ a subset of G such that every neighborhood of $1 \in G$ contains almost all elements of $\{t_i \mid i \in I\}$.*

Then there is a unique morphism $\varphi : F(I) \longrightarrow G$ such that

$$\varphi(s_i) = t_i \quad \text{for } i \in I. \tag{4.3}$$

Proof. The uniqueness of φ follows from the fact that φ is determined by (4.3).

For the proof of the existence of φ we first remark that the map in (4.3) can be extended uniquely to a homomorphism $F_I \longrightarrow G$.

Next, for $U \in \mathfrak{U}_G$ we let $\Phi(U)$ denote the kernel of the induced morphism

$$F_I \longrightarrow G/U.$$

Clearly $\Phi(U) \in \mathfrak{U}$. If Φ_U denotes the induced morphism

$$F_I/\Phi(U) \longrightarrow G/U,$$

then $\{\Phi, \Phi_U \mid U \in \mathfrak{U}_G\} : \{F_I/V \mid V \in \mathfrak{U}\} \longrightarrow \{G/U \mid U \in \mathfrak{U}_G\}$ is a morphism of projective systems. By Section 1.1, there is a corresponding morphism

$$F(I) = \varprojlim F_I/V \longrightarrow \varprojlim G/U \simeq G,$$

which maps $\prod_{V \in \mathfrak{U}} wV$ to $\prod_{U \in \mathfrak{U}_G} \varphi(w)U$ for $w \in F_I$. This morphism has the desired properties. □

For a pro-p group G, let $\operatorname{Fr}(G) = G^p[G,G]$ denote the normal subgroup of G generated by all p-th powers and commutators of elements of G. Then $\operatorname{Fr}(G)$ is the smallest normal subgroup N of G such that G/N is elementary abelian.

Theorem 4.7. *Let G_1 and G_2 be pro-p groups, and let $\varphi : G_1 \longrightarrow G_2$ be a morphism. Then φ is surjective if and only if the map $\varphi_* : G_1/\operatorname{Fr}(G_1) \longrightarrow G_2/\operatorname{Fr}(G_2)$ induced by φ is surjective.*

Proof. If $\varphi(G_1) \neq G_2$, then there exists an open normal subgroup U of G_2 such that

$$\varphi(G_1)U/U \neq G_2/U.$$

By a theorem on finite p-groups (see e.g. M. Hall [21, Ch. 12]), $\varphi(G_1)U/U$ is contained in a normal subgroup G'/U of index p in G_2/U. This implies $\operatorname{Fr}(G_2) \subseteq G'$ and

$$\varphi_*(G_1/\operatorname{Fr}(G_1)) \subseteq G'/\operatorname{Fr}(G_2),$$

i.e., φ_* is not surjective. Conversely, if φ is surjective, then clearly so is φ_*. □

We now come to the characterization of free pro-p groups.

Theorem 4.8. *Let G be a pro-p group. The following assertions are equivalent:*

(i) *G is a free pro-p group;*

(ii) *every group extension of G by a pro-p group H splits;*

(iii) *G is a projective object in the category of pro-p groups.*

Proof. (i) implies (ii): Let G be a free pro-p group with system of generators $\{s_i \mid i \in I\}$, let

$$1 \longrightarrow H \longrightarrow \overline{H} \xrightarrow{\varphi} G \longrightarrow 1 \tag{4.4}$$

be a group extension, and let $\sigma : G \longrightarrow \overline{H}$ be a continuous section. We can apply Theorem 4.6 to the free group G, to $\overline{H}$ and to the subset $\{\sigma s_i \mid i \in I\}$ of $\overline{H}$. We find that there exists a morphism $\sigma' : G \longrightarrow \overline{H}$ such that $\varphi\sigma' = \mathrm{id}$, and hence (4.4) splits.

(ii) implies (iii): Let

$$\begin{array}{ccccc} & & G & & \\ & & \downarrow & & \\ G_2 & \longrightarrow & G_1 & \longrightarrow & 1 \end{array} \tag{4.5}$$

be an exact diagram of pro-p groups. By Section 1.2 we can complete (4.5) to

$$\begin{array}{ccccc} G_2 \times_{G_1} G & \xrightarrow{\varphi} & G & & \\ {\scriptstyle \varphi_2}\downarrow & & \downarrow & & \\ G_2 & \longrightarrow & G_1 & \longrightarrow & 1 \end{array}$$

with some surjective φ. By assumption, there is a morphism $\psi : G \longrightarrow G_2 \times_{G_1} G$ such that $\varphi\psi = \mathrm{id}$. Then $\varphi_2\psi$ is the desired morphism.

(iii) implies (i): Assume that G satisfies (iii); by (4.3), $G/\operatorname{Fr}(G)$ is isomorphic to $\prod_{i\in I} \mathbb{F}_p$ for some I, and by Theorem 4.6 there is a morphism $F(I) \longrightarrow \prod_{i\in I} \mathbb{F}_p$ with kernel $\operatorname{Fr}(F(I))$. By assumption, there is a unique morphism $\varphi : G \longrightarrow F(I)$ rendering the diagram

$$\begin{array}{ccccc} & & G & & \\ & {\scriptstyle \varphi}\swarrow & \downarrow & & \\ F(I) & \longrightarrow & \prod_{i\in I} \mathbb{F}_p & \longrightarrow & 1 \end{array}$$

commutative. Since

$$\varphi^* : G/\operatorname{Fr}(G) \longrightarrow F(I)/\operatorname{Fr}(F(I))$$

is an isomorphism, φ is surjective by Theorem 4.7. Since $F(I)$ is free, there is a morphism $\psi : F(I) \longrightarrow G$ such that $\varphi\psi = \mathrm{id}$. Now ψ is surjective, hence an isomorphism, and this shows that $G \simeq F(I)$. □

We draw two conclusions from Theorem 4.8.

Theorem 4.9. *Let G be a pro-p group, I some index set, and*

$$\theta : \prod_{i\in I} \mathbb{F}_p \longrightarrow G/\operatorname{Fr}(G) \tag{4.6}$$

an epimorphism. Then there exists an epimorphism $F(I) \longrightarrow G$ that induces θ. Every system of generators of $G/\operatorname{Fr}(G)$ can be lifted to a system of generators of G.

Proof. By condition (iii) in Theorem 4.8, there is a morphism $\varphi : F(I) \longrightarrow G$ such that the diagram

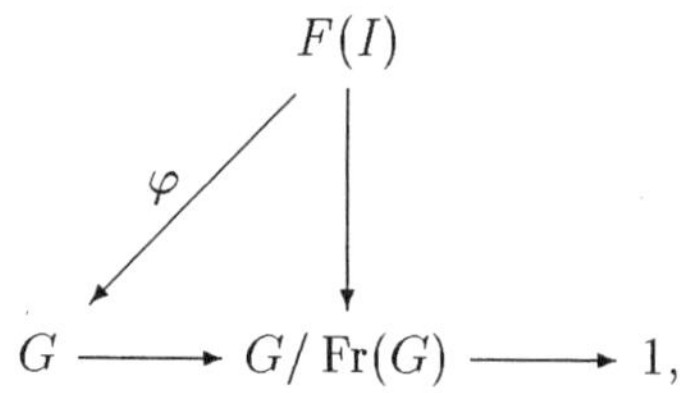

which we get using (4.6), is commutative. By Theorem 4.7, φ is surjective.

Now let $\{t_i \mid i \in I\}$ be a system of generators $G/\operatorname{Fr}(G)$. Then there is a unique epimorphism $F(I) \longrightarrow G/\operatorname{Fr}(G)$ mapping the generators S_i of $F(I)$ to the t_i for all $i \in I$. Thus $\{\varphi s_i \mid i \in I\}$ is a system of generators of G that lifts to $\{t_i \mid i \in I\}$. □

Theorem 4.10 (Burnside's Basis Theorem). *Let G be a pro-p group, and $E = \{s_i \mid i \in I\}$ a subset of G such that every neighborhood of $1 \in G$ contains almost all elements of E.*

E is a system of generators of G if and only if $\{s_i \operatorname{Fr}(G) \mid i \in I\}$ generates $G/\operatorname{Fr}(G)$.

Proof. Assume that $\{s_i \operatorname{Fr}(G) \mid i \in I\}$ generates $G/\operatorname{Fr}(G)$. There is a morphism $\varphi : F(I) \longrightarrow G$ corresponding to the system of generators E which induces an epimorphism $\varphi^* : F(I)/\operatorname{Fr}(F(I)) \longrightarrow G/\operatorname{Fr}(G)$. Then φ is surjective by Theorem 4.7, i.e., $\{s_i \mid i \in I\}$ generates G. □

4.5 Second Characterization of Free pro-p Groups

The abelian group $\mathbb{F}_p$ only has automorphisms of order coprime to p. Therefore, a pro-p group G always acts trivially on $\mathbb{F}_p$.

For the group $\mathrm{H}^n(G, \mathbb{F}_p)$, which will occur often in the following, we shall introduce the abbreviation

$$\mathrm{H}^n(G) := \mathrm{H}^n(G, \mathbb{F}_p).$$

An abelian p-group G with exponent p has $\mathrm{H}^1(G)$ as its dual group. Therefore, Theorem 4.7 can be formulated as follows:

Theorem 4.11. *Let G_1 and G_2 be pro-p groups, and let $\varphi : G_1 \longrightarrow G_2$ be a morphism. Then φ is surjective if and only if the induced map $\varphi^* : \mathrm{H}^1(G_2) \longrightarrow \mathrm{H}^1(G_1)$ is injective.*

We now give a cohomological characterization of free pro-p groups.

Theorem 4.12. *A pro-p group G is free if and only if $\mathrm{H}^2(G) = 0$.*

For the proof of Theorem 4.12 we need two lemmas:

Lemma 4.13. *Let $A \neq 1$ be a finite p-group, G a subgroup of p-power order of the automorphism group of A. Then $A^G \neq 1$.*

Proof. We decompose A into disjoint subsets of the form

$$G_a = \{ga \mid g \in G\}, \quad a \in A.$$

The number of elements in G_a is 1 if a is invariant, and a multiple of p otherwise. Thus the number of invariant elements is divisible by p, which proves the claim. □

Lemma 4.14. *Let G be a pro-p group, and let $H \neq 1$ be a normal subgroup of G. Then there exists a normal subgroup H' of G such that $H' \subset H$ and $(H : H') = p$.*

Proof. Since $H \neq 1$ there exists a proper open subgroup H'' of H. Since the topology of H is induced by the topology of G, we may assume without loss of generality that H'' is normal in G. Let H' be a maximal proper subgroup of H that is normal in G: then $(H : H') = p$. In fact, if $(H : H') > p$, then H/H' is a finite p-group on which G acts by conjugation. By Lemma 4.13, there is a subgroup H_1/H' of H/H' with order p and invariant under G, contradicting the maximality of H'. □

Proof of Theorem 4.12. By Theorem 4.8, condition (ii), and Section 3.2 we have $\mathrm{H}^2(G) = 0$ if G is a free pro-p group.

Conversely, if G is a pro-p group with $\mathrm{H}^2(G) = 0$, and if

$$\theta : \prod_I \mathbb{F}_p \longrightarrow G/\operatorname{Fr}(G)$$

is an isomorphism for some index set I, then by Theorem 4.9 there exists a morphism $\varphi : F(I) \longrightarrow G$ that induces θ. Put $H = \ker \varphi$. If we had $H \neq 1$, then by Lemma 4.14 there would exist a normal subgroup H' of $F(I)$ such that $(H : H') = p$.

Put $G' = F(I)/H'$. Since $\mathrm{H}^2(G) = 0$, the group extension

$$0 \longrightarrow \mathbb{F}_p \longrightarrow G' \longrightarrow G \longrightarrow 1$$

induced by φ splits, i.e., we have

$$G' \simeq \mathbb{F}_p \times G. \tag{4.7}$$

In the commutative diagram

$$\begin{array}{ccccc} F(I) & \longrightarrow & G' & \longrightarrow & G \\ \downarrow & & \downarrow & & \downarrow \\ \prod_I \mathbb{F}_p & \longrightarrow & G'/\operatorname{Fr}(G') & \longrightarrow & G/\operatorname{Fr}(G) \end{array}$$

the morphisms in the second row are isomorphisms, contradicting (4.7). □

Example 4.15. The quotient group of the free pro-p group $F(I)$ modulo its commutator subgroup is isomorphic to $\prod_I \mathbb{Z}_p$.

5. Cohomological Dimension

The cohomological dimension of a pro-p group equals 1 if and only if G is free. In general, the cohomological dimension is a measure of how far a pro-p group is from being free. Recall that we put $\mathrm{H}^n(G) = \mathrm{H}^n(G, \mathbb{F}_p)$.

5.1 Definition of Cohomological Dimension

Definition 5.1. A pro-p group G has *cohomological dimension* $\mathrm{cd}\, G = n$ if $n \geq 0$ is the minimal integer such that $\mathrm{H}^n(G) \neq 0$ and $\mathrm{H}^{n+1}(G) = 0$. If $\mathrm{H}^n(G) \neq 0$ for all integers $n \in \mathbb{N}$ then we put $\mathrm{cd}\, G = \infty$.

This definition is motivated by the following

Theorem 5.2. *Let G be a pro-p group with* $\mathrm{cd}\, G \leq n$. *Then* $\mathrm{H}^\nu(G, A) = 0$ *for all $\nu > n$ and all torsion modules A.*

Proof. A torsion module is the inductive limit of its finite submodules; according to Theorem 3.16 it is therefore sufficient to prove the claim for all finite G-modules A.

The module A can be written as the direct sum $A = A[p] \oplus A[p']$, where $A[p]$ and $A[p']$ are the G-submodules consisting of elements of p-power order and of order prime to p, respectively. According to Theorem 3.6, this implies

$$\mathrm{H}^m(G, A) = \mathrm{H}^m(G, A[p]) \oplus \mathrm{H}^m(G, A[p']).$$

By Theorem 3.19 we have $\mathrm{H}^m(G, A[p']) = 0$ for $m \geq 1$. It is therefore sufficient to consider finite p-primary modules A.

For such modules there is a composition series

$$0 \subseteq A_1 \subseteq A_2 \subseteq \ldots \subseteq A_s = A$$

whose factors $A_{\nu+1}/A_\nu$ are isomorphic to $\mathbb{F}_p$ by Lemma 4.13. Now induction and Theorem 3.5 imply $\mathrm{H}^{n+1}(G, A) = 0$.

By dimension shifting we get $\mathrm{H}^\nu(G, A) = 0$ for $\nu > n$. □

As a consequence of Theorem 5.2 we get

Theorem 5.3. *Let G be a pro-p group with subgroup H; then*

$$\operatorname{cd} H \leq \operatorname{cd} G.$$

Proof. According to Theorem 3.9, we have

$$\mathrm{H}^m(H, \mathbb{F}_p) = \mathrm{H}^m(G, M_G^H(\mathbb{F}_p)) = 0$$

for $m > \operatorname{cd} G$. □

5.2 Euler-Poincaré Characteristic

For a finite group A of order p^ν we put $\dim A = \nu$. The *Euler-Poincaré characteristic* $\chi(G)$ of a pro-p group G with finite cohomological dimension and with $\dim \mathrm{H}^n(G) < \infty$ for all $n \geq 1$ is defined by

$$\chi(G) = \sum_{n=0}^{\infty} (-1)^n \dim \mathrm{H}^n(G).$$

Similarly, for a G-module A of p-power order we define

$$\chi(G, A) = \sum_{n=0}^{\infty} (-1)^n \dim \mathrm{H}^n(G, A).$$

Now let

$$0 \longrightarrow A_1 \longrightarrow A_2 \longrightarrow A_3 \longrightarrow 0$$

be an exact sequence of G-modules of p-power order. The corresponding cohomology sequence yields

$$\chi(G, A_2) = \chi(G, A_1) + \chi(G, A_3).$$

For a G-module A of order p^s, an argument similar to that in the proof of Theorem 5.2 gives

$$\chi(G, A) = s \cdot \chi(G). \tag{5.1}$$

Theorem 5.4. *Let G be a pro-p group with Euler-Poincaré characteristic, and let H be a subgroup of finite index in G. Then*

$$\chi(H) = (G : H)\chi(G).$$

Proof. It follows from Theorem 3.9 and (5.1) that

$$\chi(H) = \chi(G, M_G^H(\mathbb{F}_p)) = (G : H)\chi(G).$$

□

Now we want to prove a converse to Theorem 5.4; to this end we introduce the *partial Euler-Poincaré characteristic* $\chi_n(G, A)$, which is defined for pro-p groups G with $\dim \mathrm{H}^\nu(G) < \infty$ for $0 \le \nu \le n$ by

$$\chi_n(G) = \sum_{\nu=0}^{n} (-1)^\nu \dim \mathrm{H}^\nu(G),$$

and for a G-module A of p-power order by

$$\chi_n(G, A) = \sum_{\nu=0}^{n} (-1)^\nu \dim \mathrm{H}^\nu(G, A).$$

Induction over the composition series of A together with the cohomology sequence yield

$$(-1)^n \chi_n(G, A) \le (-1)^n \chi_n(G) \dim A. \tag{5.2}$$

Theorem 5.5. *Let G be a pro-p group with partial Euler-Poincaré characteristic $\chi_n(G)$, and $\mathfrak{U}$ a system of open subgroups of G forming a neighborhood basis at $1 \in G$. If*

$$\chi_n(U) = (G : U)\chi_n(G) \tag{5.3}$$

for all $U \in \mathfrak{U}$, then $\operatorname{cd} G \le n$.

Proof. We have to show that $\mathrm{H}^{n+1}(G) = 0$.

Take some $a \in K^{n+1}(G, \mathbb{F}_p)$ representing the class $\overline{a} \in \mathrm{H}^{n+1}(G)$. There is an open subgroup $U \in \mathfrak{U}$ of G with the property that a depends only on the cosets of G/U. To the commutative diagram

$$\begin{array}{ccc} [G, \mathbb{F}_p] & \longrightarrow & [G, M_G^U(\mathbb{F}_p)] \\ & \searrow \quad \swarrow & \\ & [U, \mathbb{F}_p] & \end{array}$$

over the category $\mathcal{C}$ there corresponds the commutative diagram

$$\begin{array}{ccc} \mathrm{H}^{n+1}(G) & \xrightarrow{\ \varphi\ } & \mathrm{H}^{n+1}(G, M_G^U(\mathbb{F}_p)) \\ & {\scriptstyle \mathrm{res}} \searrow \quad \swarrow {\scriptstyle \simeq} & \\ & \mathrm{H}^{n+1}(U) & \end{array}$$

By the choice of U, we have $\mathrm{res}\,\overline{a} = 0$, hence $\varphi\overline{a} = 0$.

The cohomology sequence for the exact sequence

$$0 \longrightarrow \mathbb{F}_p \longrightarrow M_G^U(\mathbb{F}_p) \longrightarrow A \longrightarrow 0$$

gives

$$0 \longrightarrow \mathrm{H}^0(G) \longrightarrow \dots \longrightarrow \mathrm{H}^n(G, A) \longrightarrow \ker\varphi \longrightarrow 0.$$

This implies

$$\dim\ker\varphi = (-1)^n\{\chi_n(G) + \chi_n(G, A) - \chi_n(G, M_G^U(\mathbb{F}_p))\}. \tag{5.4}$$

By Theorem 3.9 and (5.3), we have

$$\chi_n(G, M_G^U(\mathbb{F}_p)) = (G : U)\chi_n(G).$$

Combined with the estimate (5.2) this implies that $\dim\ker\varphi \leq 0$, hence $\ker\varphi = 0$, which finally gives $\overline{a} = 0$. □

6. Presentation of pro-p Groups

One of the most important methods for constructing pro-p groups is the presentation by generators and relations. In particular, we shall use this method to present the Galois groups of p-extensions.

6.1 The Generator Rank

Definition 6.1. The *generator rank* $d(G)$ of a pro-p group G is defined as the dimension of $\mathrm{H}^1(G)$ as a vector space over $\mathbb{F}_p$.

This definition is motivated by the following

Theorem 6.2. *Let G be a pro-p group. The cardinality of any minimal system of generators of G equals $d(G)$.*

Proof. Let $\{s_i \mid i \in I\}$ be a minimal system of generators of G. By Theorem 4.10, $s_i \operatorname{Fr}(G)$ is a minimal system of generators of $G/\operatorname{Fr}(G)$. The map

$$1_i \longmapsto s_i \operatorname{Fr}(G), \quad i \in I,$$

induces an isomorphism $\prod_{i \in I} \mathbb{F}_p \longrightarrow G/\operatorname{Fr}(G)$. Dualizing gives

$$\mathrm{H}^1(G) \simeq \coprod_I \mathbb{F}_p$$

and so $\dim \mathrm{H}^1(G) = \#I$. □

Example 6.3. If F is a finitely generated free group, then

$$\chi(F) = 1 - d(F).$$

For a subgroup U of finite index in F, Theorem 5.4 implies

$$d(U) = (F : U)\{d(F) - 1\} + 1.$$

This is the analogue of Schreier's theorem on subgroups of discrete free groups. In general, every subgroup of finite index in a finitely generated pro-p group is finitely generated.

Example 6.4. By Theorem 6.2, all minimal systems of generators of a pro-p group have equal cardinality. A corresponding theorem for discrete groups does not hold, as the example of infinite cyclic groups shows.

6.2 Relation Systems

Definition 6.5. Let G be a pro-p group. An exact sequence

$$1 \longrightarrow R \longrightarrow F \xrightarrow{\varphi} G \longrightarrow 1, \tag{6.1}$$

where F is a free pro-p group with system of generators $\{t_i \mid i \in I\}$ is called a *presentation* of G by F. If $\{\varphi t_i \mid i \in I\}$ is a minimal system of generators of G, then the presentation is called *minimal.*

A subset $E \subseteq R$ is called a *(generating) system of relations* of G (with respect to the presentation (6.1) if

(i) R is the smallest normal subgroup of F containing E;

(ii) every open normal subgroup of R contains almost all elements of E.

We call E *minimal* if no subset of E is a system of relations for G.

Now let $\{G_i \mid i \in I\}$ be a family of pro-p groups, and $\{T_i \mid i \in I\}$ a family of normal subgroups of the G_i such that the G_i/T_i are free pro-p groups. A family $\{\varphi_i \mid i \in I\}$ of morphisms $\varphi_i : G_i \longrightarrow G$ into some pro-p group G is called *admissible* with respect to $\{T_i \mid i \in I\}$ if every open normal subgroup of G contains almost all $\varphi_i(T_i)$.

Example 6.6. *If I is finite and $T_i = G_i$, then $\{\varphi_i \mid i \in I\}$ is admissible.*

Theorem 6.7. *Let the assumptions be as in Definition 6.5. Let $\{\varphi_i \mid i \in I\}$ be admissible with respect to $\{T_i \mid i \in I\}$. Moreover, for every $i \in I$ let*

$$1 \longrightarrow R_i \longrightarrow F_i \xrightarrow{\psi_i} G_i \longrightarrow 1 \tag{6.2}$$

be a presentation of G_i.

Then there exist morphisms $\chi_i : F_i \longrightarrow F$ with restrictions $\overline{\chi}_i : R_i \longrightarrow R$ such that the diagram

$$\begin{array}{ccccc} R_i & \longrightarrow & F_i & \xrightarrow{\psi_i} & G_i \\ \downarrow{\scriptstyle \overline{\chi}_i} & & \downarrow{\scriptstyle \chi_i} & & \downarrow{\scriptstyle \varphi_i} \\ R & \longrightarrow & F & \longrightarrow & G \end{array} \tag{6.3}$$

commutes and $\{\overline{\chi}_i \mid i \in I\}$ is admissible with respect to $\{R_i \mid i \in I\}$. In this case, we speak about an admissible presentation of $\{\varphi_i \mid i \in I\}$.

Proof. Let $\sigma : G \longrightarrow F$ be a continuous section with $\sigma(1) = 1$ and let $\{t_k \mid k \in I_i\}$ be a system of generators of the free pro-p group F_i, where the images of t_k under the map $\theta_i : F_i \longrightarrow G_i/T_i$ for some subset $I_i^1 \subset I_i$ form a free system of generators of G_i/T_i and are mapped to 1 for $k \in I_i^2 = I_i \setminus I_i^1$. Then we can define a morphism $\chi_i : F_i \longrightarrow F$ by

$$\chi_i(t_k) = \sigma\varphi_i\psi_i(t_k), \quad k \in I_i. \tag{6.4}$$

Diagram (6.3) commutes for χ_i.

Let moreover N be an open normal subgroup of F. The sets $\varphi(N)$ and $\sigma^{-1}(N)$ are open neighborhoods of $1 \in G$ with $\varphi(N) \subseteq \sigma^{-1}(N)$. Let $U \subseteq \sigma^{-1}(N)$ be an open normal subgroup of G. By assumption, we have

$$\varphi_i(T_i) \subseteq U \tag{6.5}$$

for almost all $i \in I$. Let i be an index for which (6.5) holds. Then

$$\chi_i(t_k) \in \sigma(U) \subseteq N$$

for all $k \in I_i^2$ because of (6.4). But this implies $\chi_i(\ker \theta_i) \subseteq N$, and since $R_i \subseteq \ker \theta_i$, we have $\overline{\chi}_i(R_i) \subseteq R \cap N$. Since every neighborhood of $1 \in R$ contains a group $R \cap N$, this implies the claim. □

Assume that $\{\varphi_i \mid i \in I\}$ is admissible with respect to $\{T_i \mid i \in I\}$. We consider the induced maps $\varphi_i^* : \mathrm{H}^\nu(G) \longrightarrow \mathrm{H}^\nu(G_i)$. For $\alpha \in \mathrm{H}^\nu(G)$ and $\nu \geq 2$, we have $\varphi_i^*(\alpha) = 0$ for almost all i.

For suppose that f is a cocycle representing α. Then f only depends on the cosets of G modulo some open normal subgroup U of G. If i is an index such that $\varphi_i(T_i) \subseteq U$, then f induces a cocycle in $K^\nu(G_i)$ that depends only on the cosets of G_i/T_i and which therefore must be induced from a cocycle in $K^\nu(G_i/T_i)$. Since G_i/T_i is free, $\mathrm{H}^\nu(G_i/T_i)$ vanishes, hence so does $\varphi_i^* \alpha$.

Let $\{\chi_i \mid i \in I\}$ be a family of morphisms such that (6.3) commutative, and assume that $\{\overline{\chi}_i \mid i \in I\}$ is admissible with respect to $\{R_i \mid i \in I\}$. For almost all $i \in I$, the image of an $\overline{f} \in \mathrm{H}^1(R)$ under the induced map

$$\mathrm{H}^1(R) \longrightarrow \mathrm{H}^1(R_i)$$

is 0. For let U be an open normal subgroup of R on which f is constant. For almost all $i \in I$ we have a natural factorization

$$R_i \longrightarrow U \longrightarrow R$$

and therefore

$$\mathrm{H}^1(R) \longrightarrow \mathrm{H}^1(U) \longrightarrow \mathrm{H}^1(R_i).$$

This implies the claim since the image of f in $\mathrm{H}^1(U)$ vanishes. The map

$$\chi^* : \mathrm{H}^1(R) \longrightarrow \coprod_{i \in I} \mathrm{H}^1(R_i) \tag{6.6}$$

is thus well defined. We have

Theorem 6.8. *The groups $\chi_i(R_i)$, $i \in I$, generate R as a normal subgroup of F if and only if the restriction χ^* to $\mathrm{H}^1(R)^G$ is injective.*

For the proof of Theorem 6.8 we need the following

Lemma 6.9. *Let G be an arbitrary pro-p group and A a p-primary G-module. Then $A^G = 0$ implies $A = 0$.*

Proof. Each $a \in A$ generates a finite G-module A_0. If we had $A_0 \neq 0$, then Lemma 4.13 would imply $A_0^G \neq 0$ contradicting our assumption. □

We now come to the

Proof of Theorem 6.8. Let f be a homomorphism in $\mathrm{H}^1(R)^G$ with $\chi^*(f) = 0$. If R is generated as a normal subgroup of F by the groups $\chi_i(R_i)$, then $f(\chi_i(R_i)) = 0$. Since f is invariant under $h \in F$, we also have $f(h\chi_i(R)h^{-1}) = 0$ and therefore $f(R) = 0$, i.e., $f = 0$.

Conversely, assume that χ^* is injective and let R' denote the normal subgroup of F generated by the $\chi_i(R_i)$, $i \in I$. The inclusion $R' \hookrightarrow R$ induces a homomorphism $\varphi : \mathrm{H}^1(R) \longrightarrow \mathrm{H}^1(R')$ which factorizes the map (6.6):

$$\mathrm{H}^1(R) \xrightarrow{\ \varphi\ } \mathrm{H}^1(R') \longrightarrow \coprod_{i\in I} \mathrm{H}^1(R_i).$$

By assumption, $\ker\varphi$ does not contain any elements $\neq 0$ invariant under G. Now Lemma 6.9 implies that $\ker\varphi = 0$, and by Theorem 4.11 we have $R = R'$. □

We now come to the main result of this section. To formulate it, we first define the notion of a complementary set.

Definition 6.10. A subset $E \subseteq R$ is called a *complementary set* of $\{\chi_i \mid i \in I\}$ if

(i) E and $\bigcup_{i\in I} \chi_i(R_i)$ generate R as a normal subgroup of F, and

(ii) each neighborhood of $1 \in R$ contains almost all elements of E.

The set E is called *minimal* if no subset of E is a complementary set.

Theorem 6.11. *Let $\{\varphi_i \mid i \in I\}$ be a family of morphisms $\varphi_i : G_i \longrightarrow G$ admissible with respect to $\{T_i \mid i \in I\}$, and assume that we have an admissible presentation of $\{\varphi_i \mid i \in I\}$, where the presentations of G_i and G are minimal:*

$$\begin{array}{ccccccccc}
1 & \longrightarrow & R_i & \longrightarrow & F_i & \longrightarrow & G_i & \longrightarrow & 1 \\
 & & \Big\downarrow{\scriptstyle \overline{\chi}_i} & & \Big\downarrow{\scriptstyle \chi_i} & & \Big\downarrow{\scriptstyle \varphi_i} & & \\
1 & \longrightarrow & R & \longrightarrow & F & \longrightarrow & G & \longrightarrow & 1
\end{array}$$

Moreover, let E be a minimal complementary set of $\{\chi_i \mid i \in I\}$ and let φ^* be the induced map

$$\mathrm{H}^2(G) \longrightarrow \coprod_{i\in I} \mathrm{H}^2(G_i).$$

Then

$$\dim \ker\varphi^* = \#\, E.$$

Proof. If $R_j = F_j$ for $j \in E$ is the subgroup of F generated by j and if χ_j denotes the inclusion $F_j \hookrightarrow F$, then $\{\chi_j \mid j \in I \cup E\}$ is admissible, and $\bigcup_{j \in I \cup E} \chi_j(R_j)$ generates R as a normal subgroup of F. By Theorem 6.8, the induced map

$$\mathrm{H}^1(R)^G \longrightarrow \coprod_{j \in I \cup E} \mathrm{H}^1(R_j)$$

is injective. We now consider the commutative diagram

$$\begin{array}{ccc} \mathrm{H}^1(R)^G & \longrightarrow & \coprod_{j \in I \cup E} \mathrm{H}^1(R_j) \\ \uparrow & & \uparrow \\ \ker \psi^* & \longrightarrow & \coprod_{j \in E} \mathrm{H}^1(R_j) \end{array} \tag{6.7}$$

where ψ^* denotes the induced map

$$\mathrm{H}^1(R)^G \longrightarrow \coprod_{i \in I} \mathrm{H}^1(R_i)^{G_i}.$$

The second row in (6.7) is injective, and even an isomorphism by Theorem 6.8 since E is minimal.

By Theorem 3.14, we have an exact and commutative diagram

$$\begin{array}{ccccccc} & & & & \mathrm{H}^1(G) & \longrightarrow & \coprod_{i \in I} \mathrm{H}^1(G_i) \\ & & & & {\scriptstyle \mathrm{inf}}\downarrow & & \downarrow{\scriptstyle \mathrm{inf}} \\ & & & & \mathrm{H}^1(F) & \longrightarrow & \coprod_{i \in I} \mathrm{H}^1(F_i) \\ & & & & \downarrow & & \downarrow \\ 0 & \longrightarrow & \ker \psi^* & \longrightarrow & \mathrm{H}^1(R)^G & \xrightarrow{\psi^*} & \coprod_{i \in I} \mathrm{H}^1(R_i)^{G_i} \\ & & & & {\scriptstyle \mathrm{tra}}\downarrow & & \downarrow{\scriptstyle \mathrm{tra}} \\ 0 & \longrightarrow & \ker \varphi^* & \longrightarrow & \mathrm{H}^2(G) & \xrightarrow{\varphi^*} & \coprod_{i \in I} \mathrm{H}^2(G_i) \\ & & & & \downarrow & & \downarrow \\ & & & & 0 & & 0 \end{array}$$

Since the presentations of G_i and G are minimal, the inflation maps and therefore also the transgression maps are isomorphisms. Thus the induced map $\ker \psi^* \longrightarrow \ker \varphi^*$ is an isomorphism. Since

$$\dim \ker \psi^* = \dim \coprod_{j \in E} \mathrm{H}^1(R_j) = \#E,$$

this implies the claim. □

We explicitly mention two special cases of Theorem 6.11. To this end, we introduce the notion of relation ranks.

Definition 6.12. *The* relation rank *of a pro-p group G is the $\mathbb{F}_p$-dimension of* $\mathrm{H}^2(G)$.

Theorem 6.13. *The relation rank of a pro-p group equals the cardinality of any minimal system of relations.*

Theorem 6.14. *Under the assumptions of Theorem 6.11, R is generated by the subgroups $\chi_i(R_i)$, $i \in I$, if and only if φ^* is injective.*

7. Group Algebras of pro-p Groups

The results to be proved in this chapter refer again mostly to presentations of a pro-p group by generators and relations. They will be derived using the complete group algebra of a pro-p group. In this chapter, Λ will always denote a compact commutative ring with identity. For finite groups G, the additive group of the group ring $\Lambda[G]$ is isomorphic to the direct sum of $\#G$ copies of Λ, and thus inherits a topology that makes $\Lambda[G]$ into a compact group ring.

7.1 Complete Group Algebras

Let G be a profinite group. For open normal subgroups N, N' of G with $N \supseteq N'$ we can lift the natural map

$$G/N' \longrightarrow G/N$$

linearly to a homomorphism

$$\Lambda[G/N'] \longrightarrow \Lambda[G/N]$$

of group algebras. In this way we get a projective system $\{\Lambda[G/N] \mid N \in \mathfrak{U}_G\}$ of compact rings.

Definition 7.1. The *complete group algebra* $\Lambda[[G]]$ of the profinite group G over the compact ring Λ is the projective limit of the system $\{\Lambda[G/N] \mid N \in \mathfrak{U}_G\}$.

Since the algebras $\Lambda[G/N]$ are compact, so is $\Lambda[[G]]$. By

$$g \longmapsto \prod_{N \in \mathfrak{U}_G} gN,$$

G is embedded in $\Lambda[[G]]$, and the subring $\Lambda[G]$ (which is given the subspace topology) is dense in $\Lambda[[G]]$.

The other basic properties of $\Lambda[[G]]$ are expressed in the following

Theorem 7.2. (*i*) *The map $G \longrightarrow \Lambda[[G]]$ is a covariant functor from the category of profinite groups to the category of compact Λ-algebras.*

(*ii*) *Let A be a compact Λ-algebra. Every morphism $\varphi : G \longrightarrow A^{\times}$ from G into the unit group $A^{\times}$ of A can be extended uniquely to a morphism $\Lambda[[G]] \longrightarrow A$.*

(*iii*) *Let $\varphi : G \longrightarrow G'$ be a morphism of profinite groups with kernel N. The kernel of the induced morphism $\Lambda[[G]] \longrightarrow \Lambda[[G']]$ is the closed ideal $I(N)$ generated by the elements $h - 1$, $h \in N$.*

Proof. (ii) First, φ can be lifted uniquely to a continuous homomorphism $\varphi' : \Lambda[G] \longrightarrow A$. Since $\Lambda[G]$ is dense in $\Lambda[[G]]$, φ' can be lifted uniquely to $\Lambda[[G]]$.

(i) The existence of the lift of the morphism is a special case of (ii). The map $G_1 \longrightarrow G_2$ can be extended to a map $\Lambda[[G_1]] \longrightarrow \Lambda[[G_2]]$.

(iii) Without loss of generality we may assume that φ is surjective. Obviously, $I(N) \subseteq \ker \varphi$. Thus φ induces a morphism

$$\widetilde{\varphi} : \Lambda[[G]]/I(N) \longrightarrow \Lambda[[G']].$$

The restriction

$$\psi : \{G + I(N)\}/I(N) \longrightarrow G'$$

of $\widetilde{\varphi}$ is an isomorphism. By (ii), ψ^{-1} can be lifted to a morphism $\Lambda[[G']] \longrightarrow \Lambda[[G]]/I(N)$ that turns out to be the inverse of $\widetilde{\varphi}$. This implies $I(N) = \ker \varphi'$. □

Theorem 7.3. *Let Λ be a finite ring and G a profinite group. The system $\{I(N) \mid N \in \mathfrak{U}_G\}$ is a neighborhood basis at $0 \in \Lambda[[G]]$.*

Proof. This follows immediately from Theorem 1.13 and Theorem 7.2. □

7.2 Discrete and Compact G-Modules

Definition 7.4. A compact $\Lambda[[G]]$-module A is called *free* if there is an index system I such that A is isomorphic to the direct product

$$\prod_I \Lambda[[G]]$$

as a $\Lambda[[G]]$-module.

In the following we shall only consider the special case $\Lambda = \mathbb{F}_p$.

Let A be a discrete G-module with exponent p. Then A^* is a compact right $\mathbb{F}_p[G]$-module, hence we can make A^* into an $\mathbb{F}_p[[G]]$-module. Conversely, if A is a compact $\mathbb{F}_p[[G]]$-module, then A^* is a discrete G-module with exponent p.

Theorem 7.5. *Free compact $\mathbb{F}_p[[G]]$-modules and induced modules $M_G(X)$ of free discrete $\mathbb{F}_p$-modules X are dual to each other.*

Proof. By Section 1.4 it is sufficient to prove Theorem 7.5 in the special case $X = \mathbb{F}_p$. Thus we have to prove that $M_G(\mathbb{F}_p) \simeq \mathbb{F}_p[[G]]^*$.

Let χ be a character of $\mathbb{F}_p[[G]]$ and identify the subgroup of $\mathbb{Q}/\mathbb{Z}$ generated by $\frac{1}{p}$ with $\mathbb{F}_p$. Let $\varphi\chi$ denote the restriction of $\chi : \mathbb{F}_p[[G]] \longrightarrow \mathbb{F}_p$ to G. Then $\varphi\chi \in M_G(\mathbb{F}_p)$: in fact, χ is a continuous map from $\mathbb{F}_p[[G]]$ into the discrete group $\mathbb{Q}/\mathbb{Z}$, hence χ only depends on the cosets of $\mathbb{F}_p[[G]]$ modulo an ideal $I(U)$ for $U \in \mathfrak{U}_G$. But then $\varphi\chi$ only depends on the cosets of G/U, hence is continuous. It is easily verified that $\varphi : \mathbb{F}_p[[G]]^* \longrightarrow M_G(\mathbb{F}_p)$ is a homomorphism of G-modules.

Conversely, take $f \in M_G(\mathbb{F}_p)$. Then f is a function $G \longrightarrow \mathbb{F}_p$, and

$$\psi f\Big(\sum_{g \in G} x_g g\Big) = \sum_{g \in G} f(g) x_g, \quad x_g = 0 \text{ for almost all } g \in G$$

defines a continuous character of $\mathbb{F}_p[G]$ that can be lifted uniquely to $\mathbb{F}_p[[G]]$.

The map $\psi : M_G(\mathbb{F}_p) \longrightarrow \mathbb{F}_p[[G]]^*$ defined this way is obviously the inverse map of φ. This implies the claim. □

7.3 Pro-p Groups of Dimension ≤ 2

We first prove the following

Theorem 7.6. *Let*

$$1 \longrightarrow R \longrightarrow F \longrightarrow G \longrightarrow 1$$

be a presentation of the pro-p group G, and let A be a discrete torsion G-module. Let R act trivially on A and consider $\mathrm{H}^1(R, A)$ *as a G-module by Section 3.7. Then* $\mathrm{H}^1(G, \mathrm{H}^1(R, A)) \simeq \mathrm{H}^3(G, A)$.

Proof. We start by showing that the G-module $\mathrm{H}^1(R, M_G(A))$ is induced. To this end we construct an isomorphism $\varphi : \mathrm{H}^1(R, M_G(A)) \longrightarrow M_G(\mathrm{H}^1(R, A))$. The elements of $\mathrm{H}^1(R, M_G(A))$ and $M_G(\mathrm{H}^1(R, A))$ are functions with values in $R \times G$. For $f \in \mathrm{H}^1(R, M_G(A))$, we put

$$(\varphi f)(r, g) = f(\widehat{g}^{-1} r \widehat{g}, g), \quad r \in R, g \in G,$$

where $\widehat{g}$ is a representative of g in F. It can be checked immediately that φ is an isomorphism of G-modules.

Let

$$0 \longrightarrow A \longrightarrow M_G(A) \longrightarrow C \longrightarrow 0$$

be an exact sequence of G-modules. Since R is a free pro-p group and A is torsion,

$$0 \longrightarrow \mathrm{H}^1(R, A) \longrightarrow \mathrm{H}^1(R, M_G(A)) \longrightarrow \mathrm{H}^1(R, C) \longrightarrow 0$$

is also exact. From this and Theorem 3.14 we get the exact and commutative diagram

$$\begin{array}{ccccccc}
\mathrm{H}^1(F, M_G(A)) & \longrightarrow & \mathrm{H}^0(G, \mathrm{H}^1(R, M_G(A))) & \longrightarrow & 0 & & \\
\downarrow & & \downarrow & & \downarrow & & \\
\mathrm{H}^1(F, C) & \longrightarrow & \mathrm{H}^0(G, \mathrm{H}^1(R, C)) & \longrightarrow & \mathrm{H}^2(G, C) & \longrightarrow & 0 \\
\downarrow & & \downarrow & & \downarrow & & \\
0 & & \mathrm{H}^1(G, \mathrm{H}^1(R, A)) & \dashrightarrow & \mathrm{H}^3(G, A) & & \\
 & & \downarrow & & \downarrow & & \\
 & & 0 & & 0 & &
\end{array}$$

The diagram shows that the transgression $\mathrm{H}^0(G, \mathrm{H}^1(R, C)) \longrightarrow \mathrm{H}^2(G, C)$ induces an isomorphism $\mathrm{H}^1(G, \mathrm{H}^1(R, A)) \longrightarrow \mathrm{H}^3(G, A)$. □

By putting

$$\bar{r} \circ h = \overline{h^{-1} r h}, \quad r \in R, h \in F,$$

$R/[R,R]R^p$ becomes a right F-module. Since the action of F depends only on the residue classes modulo R, we can consider $R/[R,R]R^p$ as a right G-module.

We now want to prove the following theorem.

Theorem 7.7. *Let G be a pro-p group and*

$$1 \longrightarrow R \longrightarrow F \longrightarrow G \longrightarrow 1 \tag{7.1}$$

a minimal presentation of G. The following conditions are equivalent:

(*i*) *G has cohomological dimension ≤ 2;*

(*ii*) *$\mathrm{H}^1(R)$ is induced as a G-module;*

(*iii*) $\mathrm{H}^1(R) \simeq M_G\left(\prod_I \mathbb{F}_p\right)$;

(*iv*) $R/[R,R]R^p \simeq \prod_I \mathbb{F}_p[[G]]$ *as a right G-module;*

(*v*) $R/[R,R] \simeq \prod_I \mathbb{Z}_p[[G]]$ *as a right G-module.*

Here I is an index system with $\#I = \dim\ \mathrm{H}^2(G) = r(G)$.

Proof. Clearly (iii) implies (ii), and (ii) implies (i) by Theorem 7.6.

(i) implies (iii): Let $\{r_i \mid i \in I\}$ a minimal system of relations of G with respect to (7.1). We define a homomorphism $\varphi : \mathrm{H}^1(R) \longrightarrow M_G(X)$, where $X = \coprod_I \mathbb{F}_p$, by

$$(\varphi\pi)(g) = \sum_{i\in I} \pi(\widehat{g}^{-1} r_i \widehat{g}), \quad \pi \in \mathrm{H}^1(R),\ g \in G, \tag{7.2}$$

where $\widehat{g}$ is a representative of g in F.

Because of condition (ii) in the definition of relation systems, $\pi(\widehat{g}^{-1} r_i \widehat{g})$ vanishes for almost all $i \in I$, i.e., the right hand side of (7.2) is defined. Since π is continuous, there is an open normal subgroup U of F such that π only depends on the cosets of R modulo $R \cap U$. Then $\varphi\pi$ only depends on the cosets of F/U, hence is continuous. Finally it is easy to see that φ is a homomorphism of G-modules and injective. Thus we have an exact sequence

$$0 \longrightarrow \mathrm{H}^1(R) \stackrel{\varphi}{\longrightarrow} M_G(X) \longrightarrow C \longrightarrow 0.$$

The corresponding cohomology sequence yields

$$0 \longrightarrow \mathrm{H}^1(R)^G \stackrel{\varphi^*}{\longrightarrow} M_G(X)^G \longrightarrow C^G \longrightarrow \mathrm{H}^1(G, \mathrm{H}^1(R)).$$

By assumption and Theorem 7.6, $\mathrm{H}^1(G, \mathrm{H}^1(R))$ vanishes. We are now going to show that φ^* is surjective, and to this end we consider the map (6.6) in the special case where R_i is the subgroup of R generated by the r_i. Since $\{r_i \mid i \in I\}$ is a minimal system of relations, the natural map

$$\mathrm{H}^1(R)^G \longrightarrow \coprod_{i\in I} \mathrm{H}^1(R_i)$$

is an isomorphism by Theorem 6.8. Thus for every $i \in I$ there is a $\pi \in \mathrm{H}^1(R)^G$ with $\pi(r_i) = \delta_{ij}$ in $\mathbb{F}_p$ for some $j \in I$. Thus φ^* is surjective, and therefore $C^G = 0$. By Lemma 6.9, this implies $C = 0$ and thus the claim.

The equivalence of (iii) and (iv) follows from Theorem 7.5.

(iv) follows from (v) by factoring out multiples of p.

It remains to show that (iv) implies (v): Let 1_i denote the neutral element of the i-th component of $\prod_I \mathbb{Z}_p[[G]]$. The map

$$1_i \longmapsto r_i, \quad i \in I,$$

defines a homomorphism $\varphi : \prod_I \mathbb{Z}_p[[G]] \longrightarrow R/[R,R]$ of G-modules making the diagram

$$\begin{array}{ccc} \prod_I \mathbb{Z}_p[[G]] & \stackrel{\varphi}{\longrightarrow} & R/[R,R] \\ \downarrow & & \downarrow \\ \prod_I \mathbb{F}_p[[G]] & \longrightarrow & R/[R,R]R^p \end{array}$$

commutative. Consider this diagram as a diagram of pro-p groups without operators. Burnside's basis theorem (Theorem 4.10) implies that φ maps a minimal system of generators of $\prod_I \mathbb{Z}_p[[G]]$ to a minimal system of generators of $R/[R,R]$. Since R is free, φ is an isomorphism. □

7.4 Filtrations

In this section we consider filtrations of $\mathbb{F}_p[[G]]$ and G. By $I^n(G)$ we denote the closure of the n-th power of $I(G)$ in $\mathbb{F}_p[[G]]$.

Theorem 7.8. *Let G be a finitely generated pro-p group. Then the powers of $I^n(G)$ form a neighborhood basis at $0 \in \mathbb{F}_p[[G]]$.*

We first prove two lemmas.

Lemma 7.9. *Let G be a finite p-group. Then $I^n(G) = 0$ for all sufficiently large n.*

Proof. For an integer $n \geq 0$ let

$$I^n(G) = A_0 \supseteq A_1 \supseteq \cdots \supseteq A_s = \{0\}$$

be a composition series of $I^n(G)$ as a G-module. The factors $A_\nu/A_{\nu-1}$ are isomorphic to $\mathbb{F}_p$: in fact, by Lemma 4.13 $A_\nu/A_{\nu+1}$ contains a submodule isomorphic to $\mathbb{F}_p$, and since $A_\nu/A_{\nu-1}$ is irreducible, it must be isomorphic to $\mathbb{F}_p$.

Therefore,

$$ga \equiv a \pmod{A_1}$$

for all $g \in G$ and $a \in I^n(G)$, hence $(g-1)I^n(G) \subseteq A_1$. Thus $I^{n+1}(G) \subseteq A_1$, and this implies the claim. □

Lemma 7.10. *If G is a finitely generated pro-p group, then the index $(\mathbb{F}_p[[G]] : I^n(G))$ is finite for all $n \geq 1$.*

Proof. Let $s_1, \ldots, s_m$ denote the generators of G. We put

$$x_i = s_i - 1, \quad i = 1, \ldots, m, \quad I = I(G).$$

We now show by induction that I^n/I^{n+1} is finite.

I. $n = 1$. We put

$$A_1 = \mathbb{F}_p x_1 + \ldots + \mathbb{F}_p x_m.$$

$A_1 + I^2$ is a closed subgroup of I. Since

$$gg' - 1 = (g-1)(g'-1) + g - 1 + g' - 1, \tag{7.3}$$

$gg' - 1 \in A_1 + I^2$ if $g - 1, g' - 1 \in A_1 + I^2$. This implies $A_1 + I^2 = I$.
II. Assume that I^{n-1}/I^n is finite for some n, and let $I^{n-1} = A_{n-1} + I^n$ be a decomposition of abelian groups with finite A_{n-1}. Then $A_1A_{n-1} + I^{n+1}$ is a closed subgroup of I^n, and (7.3) implies

$$I^n = \langle (g-1)I^{n-1} \rangle_{\mathbb{F}_p} = \langle (g-1)A_{n-1} + I^{n+1} \rangle_{\mathbb{F}_p} = A_1A_{n-1} + I^{n-1},$$

where $\langle x_i \rangle_{\mathbb{F}_p}$ denotes the smallest closed subspace of topological $\mathbb{F}_p$ vector spaces containing the x_i. Since A_1A_{n-1} is finite, so is I^n/I^{n+1}. □

We now come to the

Proof of Thm. 7.8. By Lemma 7.10, the subsets $I^n(G)$ are open. On the other hand, the kernels of the maps

$$\varphi_U : \mathbb{F}_p[[G]] \longrightarrow \mathbb{F}_p[G/U], \quad U \in \mathfrak{U}_G,$$

form a neighborhood basis at $0 \in \mathbb{F}_p[[G]]$. Thus it remains to show that for every $U \in \mathfrak{U}_G$ there is an n such that $I^n(G/U) \subseteq \ker \varphi_U$.

By Lemma 7.9 there is an n such that $I^n(G/U) = \{0\}$. This implies $I^n(G) \subseteq \ker \varphi_U$. □

The filtration $\{I^n(G) \mid n = 1, 2, \ldots\}$ induces a filtration $\{G_n \mid n = 1, 2, \ldots\}$ of G called the *Zassenhaus filtration*:

$$G_n = \{g \mid g - 1 \in I^n(G)\};$$

G_n is a normal subgroup of G.

Theorem 7.11. *Let G be a finitely generated pro-p group; then $\{G_n \mid n = 1, 2, \ldots\}$ is a neighborhood basis at $1 \in G$. The factors G_n/G_{n+1} are in the center of G/G_{n+1}.*

Proof. By (7.3), the map

$$g \longmapsto g - 1, \quad g \in G$$

induces an injection

$$G_n/G_{n+1} \longrightarrow I^n(G)/I^{n+1}(G).$$

Thus G_n is an open normal subgroup of G.

If U, on the other hand, is an arbitrary open normal subgroup of G, then there exists an n such that $I^n(G) \subseteq \ker \varphi_U$, and this implies $G_n \subseteq U$.

In order to prove the second claim we have to show that, for all $g \in G_n$ and $h \in G$, the commutator $[g, h] = g^{-1}h^{-1}gh$ is an element of G_{n+1}. This is equivalent to $g^{-1}h^{-1}gh - 1 \in I^{n+1}(G)$, i.e., to $hg \equiv gh \pmod{I^{n+1}(G)}$. But this congruence follows immediately from (7.3). □

Theorem 7.12. *If G is a pro-p group and $g \in G_n$, then $g^p \in G_{np}$, and $[g, h] \in G_{n+1}$ for all $h \in G_n$.*

Proof. We have just proved the second claim. The first claim follows from

$$g^p - 1 = (g - 1)^p \in I^{np}(G).$$

□

We now will define a few more filtrations that we are going to need below.

Let G be a pro-p group. We inductively define subgroups $G^{(n,q)}$, where n is a natural number and $q = p^n$ or $q = 0$. For $n = 1$, we put

$$G^{(1,q)} = G.$$

Now assume that we have defined the subgroups $G^{(n,q)}$ for some $n \geq 1$; then we put

$$G^{(n+1,q)} = (G^{(n,q)})^q [G^{(n,q)}, G].$$

By definition, $G^{(n,q)}$ is a normal subgroup of G, and $G^{(n,q)}/G^{(n+1,q)}$ is in the center of $G/G^{(n+1,q)}$. We also write $G^{(n)}$ for $G^{(n,0)}$.

Theorem 7.13. *If G is a finitely generated pro-p group, then $\{G^{(n,q)} \mid n = 1, 2, \ldots\}$ is a neighborhood basis at $1 \in G$.*

Proof. We first show by induction that $(G : G^{(n,q)})$ is finite. For $n = 1$, the claim is trivial. Assume that $G/G^{(n,q)}$ is finite for some $n \geq 1$. Then $G' = G^{(n,q)}$ is a finitely generated pro-p group by Example 6.3. Moreover, $G'^q[G', G'] \subseteq G^{(n+1,q)}$. Since $G'/G'^q[G', G']$ is finite, so is $G^{(n,q)}/G^{(n+1,q)}$. This implies the claim.

For the proof of Theorem 7.13 it remains to remark that $G^{(n,q)}$ is contained in G_n by Theorem 7.12, and that $\{G_n \mid n = 1, 2, \ldots\}$ is a neighborhood basis at $1 \in G$ by Theorem 7.11. □

7.5 Computing with Commutators and Powers

We now are going to prove several rules for computing with commutators and powers. These rules are based on investigations in the group ring $\mathbb{Z}_p[[F]]$ for a finitely generated free pro-p group F.

Let C_q be the ideal in the group ring $\mathbb{Z}_p[[G]]$ generated by q and all elements of the form $g - 1$, $g \in G$, and let C_q^n denote the closure of the n-th power of C_q.

Let $g \in G^{(n,q)}$. Using induction it is easily shown that $g - 1 \in C_q^n$ and that the map

$$g \longmapsto (g - 1) + C_q^{n+1}, \quad g \in G^{(n,q)},$$

is a homomorphism. The induced map

$$\varphi_q^n : G^{(n,q)}/G^{(n+1,q)} \longrightarrow C_q^n / C_q^{n+1}$$

is a morphism of $\mathbb{Z}_p$-modules.

Theorem 7.14. *If F is a finitely generated free pro-p group, then φ_q^n is injective. Moreover, a $g \in F$ is in $F^{(n,q)}$ if and only if $g - 1 \in C_q^n$.*

Proof. Assume first that $q = 0$. Then φ_q^n is injective by M. Hall [21, Lemma 11.2.3]. (In [21, 11.2], the group ring of a free group over $\mathbb{Z}$ is considered. Since this group ring is dense in $\mathbb{Z}_p[[F]]$, the results carry over to the case considered here.) Now let $h - 1 \in C_q^n$, $h \in G^{(s)} \setminus G^{(s+1)}$. Now assume that $s < n$. Then $\varphi_q^s(hG^{(s+1)}) = 0$ contradicting the injectivity of φ_q^s.

Now let $q \neq 0$ and $h - 1 \in C_q^n$. Then there is an i with $1 \leq i < n$ such that

$$h - 1 = q^{n-1}c_i + q^{n-i-1}c_{i+1} + \ldots + c_n, \quad c_\nu \in C_0^\nu \text{ for } \nu = i, \ldots, n.$$

By the remark above we have $h \in F^{(i)}$, and since φ_0^i is a morphism of $\mathbb{Z}_p$-modules, there is an $h_i \in F^{(i)}$ such that

$$h \equiv h_i^{q^{n-i}} \pmod{F^{(i+1)}}.$$

Then $hh_1^{-q^{n-i}} - 1$ can be written in the form

$$hh_1^{-q^{n-i}} - 1 = q^{n-i-1}c'_{i+1} + \ldots + c'_n, \quad c'_\nu \in C_0^\nu \text{ for } \nu = i+1, \ldots, n,$$

and this implies that φ_q^n is injective. □

Theorem 7.15. *Let G be a pro-p group. For $a \in G^{(n,q)}$, $b \in G^{(m,q)}$ and $c \in G$ we have*

$$[a, b] \in G^{(n+m,q)}, \tag{7.4}$$
$$[a, b^q] \equiv [a, b]^q[[a, b], b]^{\binom{q}{2}} \bmod G^{(n+m+2,q)}, \tag{7.5}$$
$$[a, bc] \equiv [a, b][a, c]G^{(n+m+1,q)}, \tag{7.6}$$
$$(ac)^q \equiv a^q c^q [a, c]^{\binom{q}{2}} \bmod G^{(n+2,q)}. \tag{7.7}$$

Proof. It suffices to prove Theorem 7.15 for a finitely generated free group F. Thus we can apply Theorem 7.14: from $x = a - 1 \in C_q^n$, $y = b - 1 \in C_q^m$ we get

$$\begin{aligned} [a, b] &= a^{-1}b^{-1}ab \\ &= (1 - x + x^2 \pm \cdots)(1 - y + y^2 \pm \cdots)(1 + x)(1 + y) \\ &\equiv 0 \bmod C_q^{n+m} \end{aligned}$$

and thus $[a, b] \in F^{(n+m+1,q)}$.

The relations (7.5) – (7.7) are proved in a similar manner or directly by using the identity

$$[a, bc] = [a, c][a, b][[a, b], c]$$

and (7.4). □

7.6 Group Rings of Free pro-p Groups

In Section 4.2 we have defined the Magnus algebra $\Lambda(I)$ in the variables x_i, $i \in I$, over the ring Λ. In the special case $\Lambda = \mathbb{F}_p$ and $I = \{1, 2, \ldots, m\}$ we introduce a topology on $\Lambda(I) = \mathbb{F}_p(m)$ by giving a basis of open neighborhoods of 0. Let D^n denote the ideal of all power series in $\mathbb{F}_p(m)$ whose homogeneous components have degree at least n. Then $\{D^n \mid n = 1, 2, \ldots\}$ is an open neighborhood basis at $0 \in \mathbb{F}_p(m)$. With this topology, $\mathbb{F}_p(m)$ is the direct product of its homogeneous components, hence compact. We have

Theorem 7.16. *Let F be the free pro-p group with system of generators s_1, $\ldots$, s_m. The map*

$$x_i \longmapsto s_i, \quad i = 1, \ldots, m,$$

can be extended to an isomorphism $\mathbb{F}_p(m) \simeq \mathbb{F}_p[[F]]$.

Proof. Let $P(x_1, \ldots, x_m)$ be a power series in $\mathbb{F}_p(m)$. According to Theorem 7.8, $P(s_1 - 1, \ldots, s_m - 1)$ is a well defined element of $\mathbb{F}_p[[F]]$. The map

$$\varphi : P(x_1, \ldots, x_m) \longmapsto P(s_1 - 1, \ldots, s_m - 1)$$

is therefore a morphism of $\mathbb{F}_p(m) \longrightarrow \mathbb{F}_p[[F]]$.

On the other hand, the map

$$s_i \longmapsto 1 + x_i, \quad s_i^{-1} \longmapsto \sum_{\nu=0}^{\infty} (-x_i)^\nu$$

can be extended to a homomorphism ψ of the free group $F(m)$ generated by $s_1, \ldots, s_m$ to the unit group of $\mathbb{F}_p(m)$. Induction shows that $F(m) \cap F_n$ gets mapped to $1 + D^n$. By Theorem 7.11, this implies that ψ is continuous. Thus there is a continuous lift of ψ to F and, by Theorem 7.2.ii), to $\mathbb{F}_p[[F]]$. The map $\mathbb{F}_p[[F]] \longrightarrow \mathbb{F}_p(m)$ defined this way is obviously the inverse map of φ, i.e., φ is an isomorphism. □

In the following, we identify $\mathbb{F}_p(m)$ and $\mathbb{F}_p[[F]]$.

Theorem 7.17. *Let G be a finitely generated pro-p group with presentation*

$$1 \longrightarrow R \longrightarrow F \longrightarrow G \longrightarrow 1. \tag{7.8}$$

Let $\{r_i \mid i \in I\}$ be a system of relations with respect to the presentation (7.8). Then the kernel of the induced map $\mathbb{F}_p[[F]] \longrightarrow \mathbb{F}_p[[G]]$ is generated as an ideal of $\mathbb{F}_p[[F]]$ by $\{r_i - 1 \mid i \in I\}$.

Proof. By Theorem 7.2.(iii), the kernel of $\mathbb{F}_p[[F]] \longrightarrow \mathbb{F}_p[[G]]$ equals $I(R)$. As an ideal, $I(R)$ is generated by the elements $r_i - 1$, $i \in I$. □

7.7 The Theorem of Golod-Shafarevich

Now we turn to the main result of this chapter.

Definition 7.18. Let G be a pro-p group with finitely many generators, and let

$$1 \longrightarrow R \longrightarrow F \longrightarrow G \longrightarrow 1 \tag{7.9}$$

be a presentation of G, where F is a finitely generated free pro-p group.

The *level* of $r \in R$ is the natural number m such that $r \in F_m \setminus F_{m+1}$, where $\{F_n \mid n = 1, 2, \ldots\}$ is the Zassenhaus filtration.

We also need the following

Lemma 7.19. *Suppose that the assumptions of Definition 7.18 are satisfied. Assume moreover that S is a system of relations in G with respect to (7.9). Let S be a disjoint union of finite subsets S_n consisting of relations of level $\geq n$. (The S_n can be chosen to be finite because of our assumptions.)*

We introduce the following notation:

$$r_n = \#S_n, \; b_n = \dim\ I^n(G)/I^{n+1}(G), \; c_n = \sum_{\nu=0}^{n} b_\nu, \; r_0 = b_0 = c_0 = 1.$$

Then

$$1 \leq -c_{n-1}d + \sum_{\nu=0}^{n} c_\nu r_{n-\nu} \tag{7.10}$$

for all $n \geq 1$. Here $d = d(F)$ denotes the rank of generators of F.

Proof. Let $M = I(R)$, let L denote the closed $\mathbb{F}_p$ vector space in $\mathbb{F}_p[[F]]$ generated by the elements $r - 1$, $r \in S$, and put

$$M_n = I^n(F) \cap M, \quad L_n = I^n(F) \cap L.$$

Then

$$\mathbb{F}_p[[G]] \simeq \mathbb{F}_p[[F]]/M$$

and

$$\dim\ L/L_{n+1} \leq \sum_{\nu=1}^{n} r_\nu. \tag{7.11}$$

Moreover, let B_ν be the module of homogeneous polynomials in $x_i = s_i - 1$, $i = 1, \ldots, d$ of degree n. By Theorem 7.16, we have

$$I^n(F) = B_n \oplus I^{n+1}(F) \tag{7.12}$$

and

$$\dim\ B_n = d^n. \tag{7.13}$$

For $n = 1, 2, \ldots$ we choose subspaces $C_n \subseteq B_n$ with

$$I^n(F) \equiv M_n \oplus C_n \pmod{I^{n+1}(F)}$$

and put $C_0 = \mathbb{F}_p$. Since

$$\begin{aligned}(\mathbb{F}_p[[F]]/I^n(F))/(MI^n(F)/I^n(F)) &\simeq (\mathbb{F}_p[[F]]/M)/(MI^n(F)/M) \\ &\simeq \mathbb{F}_p[[G]]/I^n(G)\end{aligned}$$

and

$$MI^n(F)/I^n(F) \simeq M/M_n$$

we get

$$\dim\ \mathbb{F}_p[[F]]/I^n(F) = \dim\ M/M_n + \dim\ \mathbb{F}_p[[G]]/I^n(G). \tag{7.14}$$

On the other hand, we have

$$\dim\ I^n(F)/I^{n+1}(F) = \dim\ M_n/M_{n+1} + \dim\ C_n,$$

hence

$$\dim\ \mathbb{F}_p[[F]]/I^{n+1}(F) \ = \ \dim\ M/M_{n+1} + \sum_{\nu=0}^{n} \dim\ C_\nu. \tag{7.15}$$

By (7.14) and (7.15) we find

$$c_n = \sum_{\nu=0}^{n} \dim\ C_\nu, \quad b_n = \dim\ C_n.$$

After these preliminaries we now come to the heart of the proof, namely estimating dim M/M_{n+1}. Let C be the closed $\mathbb{F}_p$ vector space generated by the C_n, $n = 0, 1, \ldots$, and $A = \mathbb{F}_p[[F]]$. Then

$$A \equiv M \oplus C \pmod{I^n(F)},$$

hence

$$A = M \oplus C.$$

Moreover, we have

$$\begin{aligned}AL &= ML + CL \subseteq MAB_1 + CL = MB_1 + CL, \\ M &= ALA = ALAB_1 + AL = MB_1 + AL,\end{aligned}$$

hence

$$M \equiv MB_1 + CL \pmod{I^{n+1}(F)}.$$

This implies the estimate

$$\dim\ M/M_{n+1} \leq \dim\ MB_1/MB_1 \cap I^{n+1}(F) \\ + \dim\ CL/CL \cap I^{n+1}(F). \tag{7.16}$$

The right hand side can be transformed as follows: from the epimorphism defined by

$$\overline{m} \otimes b \longmapsto \overline{mb}, \quad m \in M, b \in B_1$$

we deduce

$$\dim\ MB_1/MB_1 \cap I^{n+1}(F) \leq d \cdot \dim\ M/M_n, \tag{7.17}$$

and the epimorphism

$$\bigoplus_{\nu=1}^{n} (C_{n-\nu} \otimes_{\mathbb{F}_p} L/L_{\nu+1}) \longrightarrow CL/CL \cap I^{n+1}(F)$$

defined by

$$c \otimes \bar{l} \longmapsto \overline{cl}, \quad c \in C_{n-\nu}, l \in L$$

implies

$$\dim\ CL/CL \cap I^{n+1}(F) \leq \sum_{\nu=1}^{n} b_{n-\nu} \dim\ L/L_{\nu+1}. \tag{7.18}$$

Collecting (7.11) and (7.15)–(7.18), we find

$$\sum_{\nu=0}^{n} d^{\nu} - c_n \leq d\Big(\sum_{\nu=0}^{n-1} d^{\nu} - c_{n-1}\Big) + \sum_{\nu=1}^{n} b_{n-\nu}\Big(\sum_{\mu=1}^{\nu} r_{\mu}\Big)$$

and after a little calculation

$$1 \leq -dc_{n-1} + \sum_{\nu=0}^{n} c_{\nu} r_{n-\nu}.$$

□

Now it is easy to prove the following theorem:

Theorem 7.20. *Let G be a finite p-group, let d, r_1, r_2, ... be defined as in Lemma 7.19, and assume that the series*

$$\varphi(t) = 1 + (r_1 - d)t + r_2 t^2 + \dots$$

converges for $0 < t < 1$. Then $\varphi(t) > 0$ for $0 < t < 1$.

Remark. Since the relation rank $r(G) = \dim\ \mathrm{H}^2(G)$ is finite, we can choose the system of relations in such a way that almost all r_n vanish.

Proof of Theorem 7.20. By Lemma 7.9, $b_n = 0$ for almost all n. Thus the series $\sum_{n=0}^{\infty} b_n t^n$ is a polynomial, and the series $\sum_{n=0}^{\infty} c_n t^n$ converges for $0 < t < 1$. Therefore we can rearrange the product:

$$\left(\sum_{n=0}^{\infty} c_n t^n\right)\left(\sum_{n=0}^{\infty} r_n t^n\right) = \sum_{n=0}^{\infty}\left(\sum_{\nu=0}^{n} c_\nu r_{n-\nu}\right) t^n.$$

Now Lemma 7.19 implies

$$\frac{1}{1-t} \leq \left(\sum_{n=0}^{\infty} c_n t^n\right)\left(\sum_{n=0}^{\infty} r_n t^n - dt\right)$$

for $0 < t < 1$, and after division by $\sum_{n=0}^{\infty} c_n t^n$

$$0 < \sum_{n=0}^{\infty} r_n t^n - dt.$$

□

We now derive a corollary from Theorem 7.20 that for $n = 2$ gives the Theorem of Golod-Shafarevich in the sharpened form of Gaschütz and Vinberg.

Theorem 7.21. *Let G be a finite p-group, and*

$$1 \longrightarrow R \longrightarrow F \longrightarrow G \longrightarrow 1$$

a minimal presentation of G with $R \subseteq F_m$ for some m, where $\{F_n \mid n = 1, 2, \ldots\}$ is the Zassenhaus filtration of F.

Then the generator rank d and the relation rank r of G are connected by the inequality

$$r > \frac{d^m}{m^m}(m-1)^{m-1}.$$

Proof. In the case at hand, $\varphi(t)$ has the form

$$\varphi(t) = -dt + rt^m.$$

Assume that $r \leq \frac{d^m}{m^m}(m-1)^{m-1}$. After plugging in $t = t_0 = (d/mr)^{1/(m-1)}$, a simple calculation shows that $\varphi(t_0) \leq 0$ in contradiction to Theorem 7.20. □

7.8 Relation Structure and Cup Product

Let G be a pro-p group with finite relation rank d, let $\{s_1, \ldots, s_d\}$ be a minimal system of generators of G, let

$$1 \longrightarrow R \longrightarrow F \longrightarrow G \longrightarrow 1 \tag{7.19}$$

be a minimal presentation of G, and let $\{r_i \mid i \in I\}$ be a minimal system of relations of G with respect to (7.19).

Then $\{s_\nu[G,G], \nu = 1, \dots, d\}$ is a minimal system of generators of $G/[G,G]$. Assume that the orders of the elements $s_\nu[G,G]$, $\nu = 1, \dots, d$ are multiples of $q = p^n$ or ∞. Then the inflation

$$\mathrm{H}^1(G, \mathbb{Z}/q) \longrightarrow \mathrm{H}^1(F, \mathbb{Z}/q),$$

where G acts trivially on $\mathbb{Z}/q$, is an isomorphism. In the following, we shall identify these groups. It follows from Theorem 3.14 and Theorem 4.12 that the transgression

$$\mathrm{H}^1(R, \mathbb{Z}/q)^G \longrightarrow \mathrm{H}^2(G, \mathbb{Z}/q)$$

is an isomorphism.

Thus we can define a homomorphism

$$\varphi_i : \mathrm{H}^2(G, \mathbb{Z}/q) \longrightarrow \mathbb{Z}/q$$

for every $i \in I$ by putting

$$\varphi_i \alpha = \operatorname{tra}^{-1} \alpha(r_i).$$

We next define a basis $\{\chi_1, \dots, \chi_d\}$ of $\mathrm{H}^1(G, \mathbb{Z}/q)$ corresponding to $\{s_1, \dots, s_d\}$ by

$$\chi_\nu(s_\mu) = \delta_{\nu\mu}, \quad \nu, \mu = 1, \dots, d.$$

Since the presentation (7.19) is minimal, we have $R \subseteq F^{(2,q)}$. For the following, we need to know more about the structure of $F^{(2,q)}/F^{(3,q)}$.

Theorem 7.22. *Every element $g \in G^{(2,q)}$ can be written in the form*

$$g = \prod_{\nu=1}^{d} s_\nu^{a_\nu q} \prod_{\nu<\mu} [s_\nu, s_\mu]^{a_{\nu\mu}} g', \quad g' \in G^{(3,q)},\ 0 \le a_\nu, a_{\nu\mu} < q. \tag{7.20}$$

If G is free, the $a_\nu, a_{\nu\mu}$ are uniquely determined.

Proof. The existence of a presentation (7.20) for a finite word in $s_1, \dots, s_d$ follows from Hall's collection process (cf. M. Hall [21, Chap. 11.1]). For an arbitrary element in $G^{(2,q)}$, (7.6) follows by writing it in the form $g_1 g'$, where g_1 is finite and $g' \in G^{(3,q)}$.

The uniqueness of the a_ν follows from the fact that $G/[G,G]G^{q^2}$ is a direct product of d cyclic groups of order q^2. In order to prove that the $a_{\nu\mu}$ are unique, we consider the cocycle

$$g_1, g_2 \longmapsto \chi_\nu(g_1)\chi_\mu(g_2).$$

Since G is free by assumption, this cocycle splits, hence there exists a continuous map $\psi_{\nu\mu} : G \longrightarrow \mathbb{Z}/q$ such that

$$\chi_\nu(g_1)\chi_\mu(g_2) = \psi_{\nu\mu}(g_1) + \psi_{\nu\mu}(g_2) - \psi_{\nu\mu}(g_1g_2). \tag{7.21}$$

Since $\psi_{\nu\mu}$ can be changed by a homomorphism $G \longrightarrow \mathbb{Z}/q$, we may assume that

$$\psi_{\nu\mu}(s_\kappa) = 0 \quad \text{for } \kappa = 1, \dots, d.$$

Then we deduce from (7.21) that there is an integer $m \geq 0$ with

$$\psi_{\nu\mu}(s_\kappa^{m+1}) = -m\delta_{\nu\kappa}\delta_{\mu\kappa} + \psi_{\nu\mu}(s_\kappa^m),$$

i.e.

$$\psi_{\nu\mu}(s_\kappa^m) = 0 \quad \text{for } \nu \neq \mu,$$

$$\psi_{\nu\mu}(s_\kappa^m) = -\binom{m}{2} \quad \text{for } \nu = \mu, \tag{7.22}$$

and since

$$\psi_{\nu\mu}(g^{-1}) + \psi_{\nu\mu}(g) = -\chi_\nu(g)\chi_\mu(g),$$

we get

$$\begin{aligned}\psi_{\nu\mu}([s_\kappa, s_\lambda]) &= \psi_{\nu\mu}(s_\kappa^{-1}) + \psi_{\nu\mu}(s_\lambda^{-1}s_\kappa s_\lambda) - \chi_\nu(s_\kappa^{-1})\chi_\mu(s_\kappa)\\ &= \psi_{\nu\mu}(s_\lambda^{-1}s_\kappa s_\lambda) \;\; = \;\; \psi_{\nu\mu}(s_\kappa s_\lambda) + \delta_{\nu\lambda}\delta_{\mu\kappa}\\ &= -\delta_{\nu\kappa}\delta_{\mu\lambda} + \delta_{\nu\lambda}\delta_{\mu\kappa},\end{aligned}$$

i.e.,

$$\psi_{\nu\mu}([s_\kappa, s_\lambda]) = \begin{cases} -1 & \text{for } \nu = \kappa, \mu = \lambda, \\ +1 & \text{for } \nu = \lambda, \mu = \kappa, \\ 0 & \text{otherwise.} \end{cases} \tag{7.23}$$

Moreover, $\psi_{\nu\mu}(G^{(3,q)}) = 0$ since $\psi_{\nu\mu}$ behaves multiplicatively when forming x^q and $[x, y]$ for $x \in G^{(2,q)}$, $y \in G$ because of (7.21).

Now we apply $\psi_{\nu\mu}$ to (7.20) and get

$$\psi_{\nu\mu}(g) = -\bar{a}_{\nu\mu} \quad \text{for } \nu < \mu,$$

which implies the uniqueness of $a_{\nu\mu}$ (the bar here and below denotes the coset modulo $q\mathbb{Z}$). □

Concerning the uniqueness of the $a_{\nu\mu}$ we could have referred to M. Hall [21, Chap. 11.1]. However, we will need the considerations above for the proof of the following theorem.

Theorem 7.23. *Let G be a pro-p group with generator rank d and minimal presentation (7.19), let q be a p-power with $R \subseteq F^{(2,q)}$, let $\{s_1, \ldots, s_d\}$ be a system of generators of F, and let $\{r_i \,|\, i \in I\}$ be a minimal system of relations of G. Assume that r_i is written according to Theorem 7.22 in the form*

$$r_i = \prod_{\kappa=1}^{d} s_\kappa^{a^{i\kappa} q} \prod_{\nu<\mu} [s_\nu, s_\mu]^{a^{i\nu\mu}} r_i', \quad r_i' \in F^{(3,q)}.$$

Then

$$\varphi_i(\chi_\nu \cup \chi_\mu) = \begin{cases} -\overline{a}^i_{\nu\mu} & \text{for } \nu < \mu, \\ -\binom{q}{2}\overline{a}^i_\nu & \text{for } \nu = \mu \end{cases}$$

for $i \in I$ and $\nu, \mu = 1, \ldots, d$. Here $\cup$ denotes the cup product given by multiplication in the ring $\mathbb{F}_p$, and φ_i, χ_ν are the functions defined above.

Proof. By construction of the transgression (see Section 3.7) and the proof of Theorem 7.22 we have

$$\varphi_i(\chi_\nu \cup \chi_\mu) = \psi_{\nu\mu}(r_i) = \begin{cases} -\overline{a}^i_{\nu\mu} & \text{for } \nu < \mu, \\ -\binom{q}{2}\overline{a}^i_\nu & \text{for } \nu = \mu. \end{cases}$$

□

Theorem 7.24. *If $B : \mathrm{H}^1(G, \mathbb{Z}/q) \longrightarrow \mathrm{H}^2(G, \mathbb{Z}/q)$ is the map induced by the exact sequence*

$$0 \longrightarrow \mathbb{Z}/q \xrightarrow{\ q\ } \mathbb{Z}/q^2 \longrightarrow \mathbb{Z}/q \longrightarrow 0,$$

then

$$\varphi_i(B\chi_\nu) = -\overline{a}^i_\nu$$

for all $i \in I$ and $\nu = 1, \ldots, d$.

Proof. For $\xi = x + q\mathbb{Z}$, $0 \le x \le q$, put $\widetilde{\xi} = x + q^2\mathbb{Z}$. Then

$$g_1, g_2 \longmapsto \frac{1}{q}\Big(\widetilde{\chi_\nu(g_1)} + \widetilde{\chi_\nu(g_2)} - \widetilde{\chi_\nu(g_1 g_2)}\Big)$$

is a cocycle representing $B\chi_\nu$. The induced cocycle with arguments in F splits. Thus there exists a continuous mapping $\psi_\nu : F \longrightarrow \mathbb{Z}/q$ such that

$$\psi_\nu(h_1) + \psi_\nu(h_2) - \psi_\nu(h_1 h_2) = \frac{1}{q}\Big(\widetilde{\chi_\nu(g_1)} + \widetilde{\chi_\nu(g_2)} - \widetilde{\chi_\nu(g_1 g_2)}\Big) \tag{7.24}$$

for all $h_1, h_2 \in F$.

Without loss of generality we can put

$$\psi_\nu(s_\kappa) = 0, \quad \kappa = 1, \ldots,$$

If $h_1 \in \ker \chi_\nu$ or $h_2 \in \ker \chi_\nu$ then the right hand side of (7.24) vanishes. Moreover,

$$\psi_\nu(h^{-1}) = -\psi_\nu(h) + \frac{1}{q}(\widetilde{\chi_\nu(h)} + \widetilde{\chi_\nu(h^{-1})}), \quad \in F$$

and

$$\begin{aligned}
\psi_\nu([h_1, h_2]) &= -\psi_\nu(h_2^{-1} h_1 h_2) \\
&= -\psi_\nu(h_1) + \psi_\nu(h_2^{-1}) + \psi_\nu(h_1 h_2) \\
&\qquad - \frac{1}{q}\Big(\widetilde{\chi_\nu(h_2^{-1})} + \widetilde{\chi_\nu(h_1 h_2)} - \widetilde{\chi_\nu(h_1)}\Big) \\
&= -\psi_\nu(h_1) - \psi_\nu(h_2) \\
&\qquad + \frac{1}{q}\Big(\widetilde{\chi_\nu(h_2)} + \widetilde{\chi_\nu(h_1)} - \widetilde{\chi_\nu(h_1 h_2)}\Big) + \psi_\nu(h_1 h_2) \;=\; 0
\end{aligned}$$

Thus ψ_ν vanishes on the subgroup of F generated by $s_1, \ldots, s_{\nu-1}, s_{\nu+1}, \ldots, s_d$, $[F, F]$. Finally

$$\psi_\nu(s_\nu^{aq}) \;=\; -\overline{a}.$$

This implies

$$\varphi_i(B\chi_\nu) \;=\; \psi_\nu(r_i) = -\overline{a}_\nu^i.$$

$\square$

8. Results from Algebraic Number Theory

In this chapter we formulate the theorems from class field theory for finite extensions that we shall need in the following, and we will transfer them to infinite extension as far as is necessary.

8.1 Algebraic Number Theory for Infinite Extensions

Let K/k be a (finite or infinite) algebraic extension. A place $\mathfrak{P}$ of K is an equivalence class of nontrivial valuations of K. If $\mathfrak{p}$ is the restriction of $\mathfrak{P}$ to k, we write $\mathfrak{P} \mid \mathfrak{p}$. The *algebraic completion* $K_{\mathfrak{P}}$ of K with respect to $\mathfrak{P}$ (and k) is the inductive limit $\varinjlim L_{\mathfrak{q}}$, where L runs through all finite subextensions of K/k, where $\mathfrak{q}$ is the restriction of $\mathfrak{P}$ to L, and where $L_{\mathfrak{q}}$ is the usual $\mathfrak{q}$-adic completion of L. There is a canonical embedding $K \longrightarrow K_{\mathfrak{P}}$. A place of K is called finite or infinite according as it consists of non-archimedean or archimedean valuations. A finite place is called a prime divisor. An infinite place is called *real* or *complex* according as the corresponding completion is $\mathbb{R}$ or $\mathbb{C}$. The residue class field of k with respect to $\mathfrak{p}$ will be denoted by $\mathfrak{K}_{\mathfrak{p}}$. Let L be an arbitrary intermediate field of K/k, and let $v_{\mathfrak{P}}$ be an additive valuation of K with restrictions $v_{\mathfrak{q}}$ and $v_{\mathfrak{p}}$ on L and k, respectively. The place $\mathfrak{P}$ is called *unramified* in L/k if the valuation groups of $v_{\mathfrak{q}}$ and $v_{\mathfrak{p}}$ coincide, i.e., if $v_{\mathfrak{p}}(k^{\times}) = v_{\mathfrak{q}}(L^{\times})$.

Clearly $\mathfrak{P}$ is unramified in K/k if and only if $\mathfrak{P}$ is unramified in every finite subextension of K/k. Therefore, the theorem that a prime $\mathfrak{P}$ of the compositum KK_1/k is unramified if and only if its restrictions to K and K_1 are unramified over k continues to hold for infinite extensions.

An infinite prime $\mathfrak{P} \mid \mathfrak{p}$ is called *unramified* in K/k if $K_{\mathfrak{P}} = k_{\mathfrak{p}}$.

8.2 Normal Extensions

Let K/k be a normal extension and $v_{\mathfrak{P}}$ an additive valuation of K. The *decomposition group* $\mathcal{Z}_{\mathfrak{P}}(K/k)$ and the *inertia group* $\mathcal{T}_{\mathfrak{P}}(K/k)$ of $\mathfrak{P}$ are defined as follows: an element $g \in \mathrm{Gal}(K/k)$ belongs to $\mathcal{Z}_{\mathfrak{P}}$ if

$$v_{\mathfrak{P}}(g\alpha) = v_{\mathfrak{P}}(\alpha)$$

for all $\alpha \in K$.

An automorphism $g \in \mathrm{Gal}(K/k)$ belongs to $\mathcal{T}_{\mathfrak{P}}$ if

$$v_{\mathfrak{P}}(g\alpha - \alpha) > 0$$

for all $\alpha \in K$ with $v_{\mathfrak{P}}(\alpha) \geq 0$. The inertia group $\mathcal{T}_{\mathfrak{P}}$ is normal in the decomposition group $\mathcal{Z}_{\mathfrak{P}}$.

Theorem 8.1. *Let L/k be a subextension of the normal extension K/k, and let $\mathfrak{P}$ be a finite place in K. Then*

$$\begin{aligned} \mathcal{Z}_{\mathfrak{P}}(K/L) &= \mathrm{Gal}(K/L) \cap \mathcal{Z}_{\mathfrak{P}}(K/k), \\ \mathcal{T}_{\mathfrak{P}}(K/L) &= \mathrm{Gal}(K/L) \cap \mathcal{T}_{\mathfrak{P}}(K/k). \end{aligned} \tag{8.1}$$

If L/k is normal and if $\mathfrak{q}$ denotes the restriction of $\mathfrak{P}$ to L, then

$$\mathcal{Z}_{\mathfrak{q}}(L/k) = \mathrm{Gal}(K/L)\mathcal{Z}_{\mathfrak{P}}(K/k)/Gal(K/L), \tag{8.2}$$

$$\mathcal{T}_{\mathfrak{q}}(L/k) = \mathrm{Gal}(K/L)\mathcal{T}_{\mathfrak{P}}(K/k)/Gal(K/L). \tag{8.3}$$

Proof. (8.1) follows immediately from the definition. (8.2) and (8.3) follow from the corresponding claims for finite extensions. □

It follows from Theorem 8.1 that $\mathcal{Z}_{\mathfrak{P}}(K/k) \simeq \varprojlim \mathcal{Z}_{\mathfrak{q}}(L/k)$ and $\mathcal{T}_{\mathfrak{P}}(K/k) \simeq \varprojlim \mathcal{T}_{\mathfrak{q}}(L/k)$, where the limit is over all finite normal subextensions L/k of K/k, and where $\mathfrak{q}$ is the restriction of $\mathfrak{P}$ to L.

Let $\mathfrak{p}$ denote the restriction of $\mathfrak{P}$ to k. The canonical embedding $K \hookrightarrow K_{\mathfrak{P}}$ induces an injection

$$\varphi_{\mathfrak{P}} : \mathrm{Gal}(K_{\mathfrak{P}}/k_{\mathfrak{p}}) \longrightarrow \mathrm{Gal}(K/k).$$

Theorem 8.2. *The image of the injection $\varphi_{\mathfrak{P}}$ equals $\mathcal{Z}_{\mathfrak{P}}$.*

Proof. Theorem 8.2 can be proved easily by constructing the inverse map using Cauchy sequences. □

For infinite primes we define $\mathcal{Z}_{\mathfrak{P}} = \mathcal{T}_{\mathfrak{P}}$ as the image of $\varphi_{\mathfrak{P}}$.

Theorem 8.3. *The fixed field of $\mathcal{T}_{\mathfrak{P}}$ is the maximal extension of k in which $\mathfrak{P}$ is unramified.*

Proof. If $K(\mathcal{T}_{\mathfrak{P}})$ is the fixed field of $\mathcal{T}_{\mathfrak{P}}$, then

$$K(\mathcal{T}_{\mathfrak{P}}) = \bigcup_{\mathfrak{q}} K(\mathcal{T}_{\mathfrak{q}}),$$

where the union is over all restrictions $\mathfrak{q}$ of $\mathfrak{P}$ to finite normal subextensions of K/k. Thus if $K(\mathcal{T}_{\mathfrak{q}})/k$ is unramified, then so is $K(\mathcal{T}_{\mathfrak{P}})/k$.

If, on the other hand, K'/k is a subextension of K/k in which $\mathfrak{P}$ is unramified, then every finite subextension of K'/k is contained in an extension $K(\mathcal{T}_{\mathfrak{q}})/k$, and therefore $K' \subseteq K(\mathcal{T}_{\mathfrak{P}})$. □

8.3 The Frobenius Automorphism

We now consider fields k and prime divisors $\mathfrak{p}$ of k for which the residue class field $\mathfrak{K}_\mathfrak{p}$ is finite.

If K/k is a normal finite extension and $\mathfrak{P}$ an extension of $\mathfrak{p}$ to K, then the natural morphism

$$\mathcal{Z}_\mathfrak{P} \longrightarrow \operatorname{Gal}(\mathfrak{K}_\mathfrak{P}/\mathfrak{K}_\mathfrak{p}) \tag{8.4}$$

is surjective with kernel $\mathcal{T}_\mathfrak{P}$. By Section 2.2, this fact transfers immediately to infinite extensions.

For finite $\mathfrak{K}_\mathfrak{p}$, $\operatorname{Gal}(\mathfrak{K}_\mathfrak{P}/\mathfrak{K}_\mathfrak{p})$ is canonically isomorphic to a factor group of the total completion of $\mathbb{Z}$. Let $N\mathfrak{p}$ denote the number of elements of $\mathfrak{K}_\mathfrak{p}$. Raising to the $N\mathfrak{p}$-th power is an automorphism of $\mathfrak{K}_\mathfrak{P}/\mathfrak{K}_\mathfrak{p}$ that generates $\operatorname{Gal}(\mathfrak{K}_\mathfrak{P}/\mathfrak{K}_\mathfrak{p})$. The preimage $\sigma_\mathfrak{P}(K/k)$ of this automorphism under (8.4) is called the *Frobenius automorphism* of $\mathfrak{P}$ with respect to $K(\mathcal{T}_\mathfrak{P})/k$.

Let $\mathfrak{P}$ be unramified in K/k, L/k a normal subextension, and $\mathfrak{q}$ the restriction of $\mathfrak{P}$ to L. Then $\sigma_\mathfrak{q}(L/k)$ is the restriction of $\sigma_\mathfrak{P}(K/k)$ to L.

8.4 Local and Global Fields

In the following, k always denotes a local or global field. Here, a *local field* is an algebraic extension of the field $\mathbb{Q}_p$ of rational p-adic numbers or of the field $\mathbb{F}_p((x))$ of formal Laurent series of one variable x over the field with p elements. If the extension is finite, we call k a *local field of finite type.* A *global field* is an algebraic extension of the field $\mathbb{Q}$ of rational numbers or of the field $\mathbb{F}_p(x)$ of rational functions of one variable x over the field with p elements. If the extension is finite, we call k a *global field of finite type.* Thus the local fields are the completions of global fields with respect to their prime divisors. We shall treat the fields $\mathbb{R}$ and $\mathbb{C}$ occasionally as local fields.

The divisor group $D(L)$ of a global field L of finite type is the group of formal linear combinations of the prime divisors $\mathfrak{q}$ of L with coefficients in $\mathbb{Z}$.

Let $v_\mathfrak{q}$ denote the additive valuation of L with value group $\mathbb{Z}$. Then

$$\alpha \longmapsto \sum_\mathfrak{q} v_\mathfrak{q}(\alpha)\mathfrak{q}, \quad \alpha \in L^\times,$$

is a homomorphism $(\ ) : L^\times \longrightarrow D(L)$. The cokernel of $(\ )$ is called the *class group* of L and will be denoted by $\operatorname{Cl}(L)$. Now let L' be a global field contained in L. *There exists a unique monomorphism $D(L') \longrightarrow D(L)$ compatible with the embedding $L' \subseteq L$.* The commutative diagram

$$\begin{array}{ccccc} L'^\times & \longrightarrow & D(L') & \longrightarrow & \operatorname{Cl}(L') \\ \downarrow & & \downarrow & & \downarrow \\ L^\times & \longrightarrow & D(L) & \longrightarrow & \operatorname{Cl}(L) \end{array}$$

allows us to define the group $D(K)$ of divisors and the class group $\mathrm{Cl}(K)$ of a general global field as the inductive limit of the corresponding groups for the subfields of K of finite type.

For every local or global field k of finite type there is a locally compact group $A(k)$, the *formation module.*

If k is a local field, then $A(k) = k^\times$ is the multiplicative group of k.

If k is a global field, we first define the group J_k of ideles. As an abstract group, J_k is the subgroup of $\prod_\mathfrak{p} k_\mathfrak{p}^\times$ (the product here and below is taken over all places $\mathfrak{p}$ of k) consisting of all elements whose components are units almost everywhere. For finite $\mathfrak{p}$, we let $E_\mathfrak{p}$ be the unit group, and for infinite $\mathfrak{p}$ we take $E_\mathfrak{p} = k_\mathfrak{p}^\times$. The idele group J_k contains the subgroup $\prod_\mathfrak{p} E_\mathfrak{p}$. We transfer the product topology of $\prod_\mathfrak{p} E_\mathfrak{p}$ to J_k, i.e., the basis of open neighborhoods at $1 \in \prod_\mathfrak{p} E_\mathfrak{p}$ is a neighborhood basis at $1 \in J_k$. By Tychonov's theorem, $\prod_\mathfrak{p} E_\mathfrak{p}$ and J_k are locally compact. To each idele $\prod_\mathfrak{p} \alpha_\mathfrak{p} \in J = J_k$ we can attach a divisor by putting

$$\prod_\mathfrak{p} \alpha_\mathfrak{p} \longmapsto \prod_{\mathfrak{p}\text{ finite}} \mathfrak{p}^{v_\mathfrak{p}(\alpha_\mathfrak{p})}.$$

The degree of the divisor is also called the *degree of the idele.*

Let K be another global field of finite type and $\varphi : k \longrightarrow K$ a monomorphism. Then φ induces a surjective map φ' from the set of primes of K to the set of primes of k:

$$v_{\varphi'(\mathfrak{P})}(\alpha) = v_\mathfrak{P}(\varphi(\alpha)), \quad \alpha \in k,$$

and a monomorphism $\varphi_* : k_{\varphi'(\mathfrak{P})} \longrightarrow K_\mathfrak{P}$.

Accordingly,

$$\prod_\mathfrak{p} \alpha_\mathfrak{p} \longmapsto \prod_\mathfrak{P} \varphi_*(\alpha_{\varphi'(\mathfrak{P})})$$

defines a monomorphism $J_k \longrightarrow J_K$ denoted also by φ. For $k \subseteq K$, we identify J_k with the image of $J_k \hookrightarrow J_K$. If K/k is normal, then $\mathrm{Gal}(K/k)$ acts on J_K, and we have $J(K)^{\mathrm{Gal}(K/k)} = J_k$.

We can embed $k^\times$ into J_k by mapping $\alpha \in k^\times$ to the idele (α) whose $\mathfrak{p}$-component is the image of α under the canonical embedding $k \hookrightarrow k_\mathfrak{p}$. Such ideles are called *principal ideles.*

Now we define $A(k)$ as the quotient group $C_k = J_k/k^\times$. This group C_k is called the *idele class group* of k. From our considerations above, a monomorphism $\varphi : k \longrightarrow K$ induces a monomorphism $C_k \longrightarrow C_K$. For $k \subseteq K$ we identify C_k with its image in C_K. For normal extensions K/k, $G = \mathrm{Gal}(K/k)$ acts on C_K, and we have $C(K)^G = C_k$. In fact, the exact sequence

$$1 \longrightarrow K^\times \longrightarrow J_K \longrightarrow C_K \longrightarrow 1$$

induces the cohomology sequence

$$1 \longrightarrow k^\times \longrightarrow J_k \longrightarrow C(K)^G \longrightarrow \mathrm{H}^1(G, K^\times).$$

By Theorem 3.18, $\mathrm{H}^1(G, K^\times) = 0$, hence $C(K)^G = C_k$.

Now let k be a local or global field of finite type, L a finite separable extension of k, and K a normal finite extension of k containing L. Let M be the set of isomorphisms $L \longrightarrow K$ leaving k fixed. We define the norm $N_{L/K}a$ for $a \in A(L)$ by

$$N_{L/K}a = \prod_{g \in M} ga;$$

$N_{L/K}a$ is invariant under $\mathrm{Gal}(K/k)$, hence is an element of $A(k)$.

Let k be a global field of finite type and $\mathfrak{p}$ a prime in k. We embed $k_\mathfrak{p}^\times$ into J_k by sending $\alpha \in k_\mathfrak{p}^\times$ to the idele whose $\mathfrak{p}$-component is α and whose other components are 1. We denote this idele by α. The corresponding map $k_\mathfrak{p}^\times \longrightarrow C_k$ is also injective.

Example 8.4. We have $C_\mathbb{Q} \simeq \mathbb{R}_+ \times \prod_p E_p$, where the product is over all primes, $\mathbb{R}_+$ is the multiplicative group of positive real numbers, and where E_p is the group of rational p-adic units.

8.5 The Multiplicative Group of a Local Field

Let k be a local field of finite type, $\mathfrak{p}$ the maximal ideal of k, and q the characteristic of its residue class field. The topology of k is defined by the neighborhood basis $\{E_n \,|\, n = 1, 2, \ldots\}$ at 1, where $E_n = \{\alpha \,|\, \alpha \in k^\times, \alpha \equiv 1\ (\mathfrak{p}^n)\}$ the group of *principal units* of level n. *The groups E_n are open, closed and compact, E_n/E_{n+1} is isomorphic to the finite q-group $\mathfrak{K}_\mathfrak{p}^+$, and therefore $E_n = \varinjlim_\nu E_n/E_{n+\nu}$ is a pro-q group. In particular, E_n is a $\mathbb{Z}_q$-module by Section 4.3.*

Let π be a prime element (uniformizer) of k, i.e., a generator of the maximal ideal $\mathfrak{p}$ in k. *The cyclic group generated by π is closed in $k^\times$, i.e, is a discrete subgroup. Since every $\alpha \in k^\times$ can be written uniquely in the form*

$$\alpha = \pi^{v_\mathfrak{p}(\alpha)}\zeta \cdot \varepsilon,$$

where ζ is a root of unity of order coprime to q and $\varepsilon \in E_1$, we have a direct product decomposition

$$k^\times = \langle \pi \rangle \times H \times E_1,$$

where H is the group of roots of unity in k of order coprime to q; this group has order $N(\mathfrak{p}) - 1 = \#\mathfrak{K}_\mathfrak{p} - 1$.

The structure of E_1 in the cases where $\chi(k) = q$ and $\chi(k) = 0$ is very different and is described by a theorem of Hensel. *In the first case, E_1 is the direct product of countably many copies of $\mathbb{Z}_q$. In the second case, E_1 is the direct product of the finite group of roots of unity of q-power order and of $(k : \mathbb{Q}_q)$ copies of $\mathbb{Z}_q$. In the latter case, $E_n \subseteq E_1^{q^s}$ for sufficiently large s.*

We now apply this to study the structure of tamely ramified extensions of k. Let K/k be a finite, tamely ramified extension, i.e. assume that the ramification index of K/k is coprime to q. *The extension K/k contains the unramified extension T_k/k of inertia degree f, where T_k is the compositum of k and the field of $N(\mathfrak{p})^f - 1$-th roots of unity. K/T_k is totally ramified of degree e.*

Let $\widehat{\pi}$ be a prime element of K, Then we can write

$$\widehat{\pi}^e = \pi \cdot \zeta \cdot \varepsilon,$$

where $\zeta \in T_k$ is a root of unity of order coprime to q, and ε a principal unit. Since $(e, q) = 1$, we have $1/e \in \mathbb{Z}_q$, hence ε is an e-th power in K. Therefore $\sqrt[e]{\pi\zeta} \in K$ and hence $K = T_k(\sqrt[e]{\pi\zeta})$.

The extension K/k is normal if and only if the e-th roots of unity and $\sqrt[e]{\pi\zeta^{N(\mathfrak{p})}}$ lie in K.

Now let K/k be a p-extension, i.e., a normal extension of p-power degree, $p \neq q$. If k and hence K does not contain the p-th roots of unity, then $e = 1$, i.e., K/k is unramified. Since the group of units in k with order coprime to q has order $N(\mathfrak{p}) - 1$, k contains the p-th roots of unity if and only if $N(\mathfrak{p}) \equiv 1 \bmod p$.

Now let k be a field that is complete with respect to an infinite prime $\mathfrak{p}$, i.e., let k be the field of real or complex numbers. In order to get results that are formally analogous to those above, we put

$$\alpha \equiv \beta \pmod{\mathfrak{p}} \quad \text{for } \alpha, \beta \in k \text{ if } \begin{cases} \alpha\beta^{-1} > 0 & \text{for } k = \mathbb{R}, \\ \alpha\beta^{-1} \neq 0 & \text{for } k = \mathbb{C}. \end{cases}$$

Every $\alpha \equiv 1 \bmod \mathfrak{p}$ is an n-th power for any natural number n.

8.6 Finite Class Field Theory

Let k be a local or global field of finite type. Class field theory gives a correspondence between abelian extensions of k and subgroups of $A(k)$.

There is a bijection between the (closed) subgroups U of finite index in $A(k)$ and the finite abelian extensions K/k, and we have $\mathrm{Gal}(K/k) \simeq A(k)/U$.

This isomorphism is induced by the reciprocity map

$$A(k) \longrightarrow \mathrm{Gal}(K/k) : \alpha \longmapsto (\alpha, K/k).$$

The kernel U of this homomorphism equals the norm group $N_{K/k}A(K)$.

The *Artin symbol* $(\alpha, K/k)$ satisfies the following transformation rules:

(i) *If k'/k is an arbitrary finite extension, then*

$$(N_{k'/k}\alpha, K/k) = (\alpha, Kk'/k')|_k, \quad \alpha \in A(k').$$

(ii) *If K'/k is a finite normal extension, K/k the maximal abelian subextension of K'/k, and k' a subfield of K'/k such that K'/k' is abelian, then*

$$(\alpha, K'/k') = \mathrm{Ver}\,(\alpha, K/k), \quad \alpha \in A(k),$$

where Ver *denotes the transfer* $\mathrm{Gal}\,(K'/k) \longrightarrow \mathrm{Gal}\,(K'/k')$ *(cf. Section 3.6).*

(iii) *If K'/k is a normal subextension of K/k, then*

$$(\alpha, K'/k) = (\alpha, K/k)|_{K'}, \quad \alpha \in A(k).$$

(iv) *If $s : K \longrightarrow sK$ is an isomorphism, then*

$$(s\alpha, sK/sk) = s(\alpha, K/k)s^{-1}, \quad \alpha \in A(k).$$

Here, σ_F denotes the restriction of an automorphism σ to a subfield F.

For a local field k, the Artin symbol maps the unit group of k to the inertia subgroup of K/k. If K/k is unramified, then

$$(\alpha, K/k) = \sigma(K/k)^{v(\alpha)} \quad \text{for } \alpha \in k, \tag{8.5}$$

where $\sigma(K/k)$ denotes the Frobenius automorphism of K/k and $v(\alpha)$ the normalized additive valuation of k with value group $\mathbb{Z}$.

The *conductor* of a subgroup U of finite index in $k^\times$ is the smallest power $\mathfrak{p}^n$ such that $E_n \subseteq U$.

Now let k be a global field of finite type and $\mathfrak{P}$ a prime of K whose restriction to k is $\mathfrak{p}$. Then the diagram

$$\begin{array}{ccc} k_\mathfrak{p}^\times & \longrightarrow & C_k \\ \downarrow & & \downarrow \\ \mathrm{Gal}\,(K_\mathfrak{P}/k_\mathfrak{p}) & \longrightarrow & \mathrm{Gal}\,(K/k) \end{array}$$

is commutative; here the horizontal arrows are the embeddings defined above, and the vertical arrows are the reciprocity maps. For unramified $\mathfrak{P}|\mathfrak{p}$, (8.5) implies

$$(\alpha, K/k) = \sigma_\mathfrak{P}(K/k)^{v_\mathfrak{p}(\alpha)} \quad \text{for } \alpha \in k_\mathfrak{p}. \tag{8.6}$$

Let U be a subgroup of finite index in C_k. The preimage $U_\mathfrak{p}$ of U under the embedding $k_\mathfrak{p}^\times \hookrightarrow C_k$ has finite index in $k_\mathfrak{p}^\times$. For finite primes $\mathfrak{p}$, the conductor of $U_\mathfrak{p}$ is called the $\mathfrak{p}$-conductor of U. For infinite primes $\mathfrak{p}$, the $\mathfrak{p}$-conductor of U is 1 or $\mathfrak{p}$ according as $U_\mathfrak{p} = k_\mathfrak{p}^\times$ or $U_\mathfrak{p} \neq k_\mathfrak{p}^\times$. The last case occurs only if $k_\mathfrak{p}^\times = \mathbb{R}^\times$ and $U_\mathfrak{p} = \mathbb{R}_+$.

The *conductor* of U is defined as the product of the $\mathfrak{p}$-conductors. Since U is an open subgroup of C_k, the $\mathfrak{p}$-conductor of U is 1 for almost all $\mathfrak{p}$. The finite part of the conductor of U is therefore an integral ideal of k.

If K/k is the abelian extension corresponding to U, then the finite part of the conductor of U divides the discriminant of K/k.

This theorem, together with property (8.6), uniquely determines the homomorphism $(\ ,K/k): C_k \longrightarrow \mathrm{Gal}(K/k)$. In fact, the image of an idele class whose components equal 1 at every ramified place is determined by (8.6): let $\prod \alpha_p$ be an arbitrary idele, let $d_\mathfrak{p}$ denote the $\mathfrak{p}$-component of the discriminant of K/k if $\mathfrak{p}$ is finite, and put $d_\mathfrak{p} = \mathfrak{p}$ if $\mathfrak{p}$ is infinite. By the approximation theorem of valuation theory, there exists an $\alpha \in k$ such that

$$\alpha_\mathfrak{p}\alpha^{-1} \equiv 1 \pmod{d_\mathfrak{p}}$$

for all ramified primes (an infinite place $\mathfrak{P}$ of K is called *ramified* if $K_\mathfrak{P} \neq k_\mathfrak{p}$). For ramified primes $\mathfrak{p}$ we then get $(\alpha_\mathfrak{p}\alpha^{-1}, K/k) = 1$, and for the other primes $(\alpha_\mathfrak{p}\alpha^{-1}, K/k)$ is determined by (8.6). This implies the claim.

8.7 Infinite Class Field Theory

Let k be a local or global field of finite type, K/k an abelian extension, and $a \in A(k)$. Then

$$a \longmapsto (a, K/k) = \varprojlim (a, K_\nu/k),$$

where K_ν/k runs through all finite subextensions of k/k, defines a map $A(k) \longrightarrow \mathrm{Gal}(K/k)$ whose image is everywhere dense by the transformation rule (iii) in Section 8.6. Since the subgroups of finite index in $A(k)$ and $\mathrm{Gal}(K/k)$ correspond to each other biuniquely, $\mathrm{Gal}(K/k)$ is isomorphic to the completion $\widehat{A(k)}$ of $A(k)$. The homomorphism $(\,\cdot\,, K/k)$ obviously satisfies the transformation rules (i), (iii), (iv).

Now let K be the maximal abelian extension of k. We want to study the kernel and the image of $(\,\cdot\,, K/k)$ in each in the different cases.

First let k be a local field. Then $A(k) \simeq \mathbb{Z} \times E$ by Section 8.5, where E is the unit group of k. Thus

$$\mathrm{Gal}(K/k) \simeq \widehat{A(k)} \simeq \widehat{\mathbb{Z}} \times E.$$

Now let k be a global field. If $\mathfrak{p}$ is a finite prime in k, then the valuation whose value at a uniformizer is $N(\mathfrak{p})^{-1}$ is called the *normalized valuation* $\varphi_\mathfrak{p}$; for an infinite prime, we take $\varphi_\mathfrak{p}$ as the absolute value or its square according as $k_\mathfrak{p} = \mathbb{R}$ or $k_\mathfrak{p} = \mathbb{C}$. For every $\alpha \in k$ we have $\prod_\mathfrak{p} \varphi_\mathfrak{p}(\alpha) = 1$. Therefore

$$\prod_\mathfrak{p} \alpha_\mathfrak{p} \longmapsto \prod_\mathfrak{p} \varphi_\mathfrak{p}(\alpha_\mathfrak{p})$$

defines a homomorphism $C_k \longrightarrow \mathbb{R}_+$, whose kernel we shall denote by $C_0(k)$. The group $C_0(k)$ is compact.

For a function field k, $C_0(k)$ is totally disconnected, and $C_k \simeq \mathbb{Z} \times C_0(k)$. Thus

$$\mathrm{Gal}\,(K/k) \simeq \widehat{A(k)} \simeq \widehat{\mathbb{Z}} \times C_0(k).$$

For an algebraic number field k, on the other hand, $(\,\cdot\,, K/k)$ is not injective. Its kernel $D(k)$ consists of all divisible elements, and $C_k \simeq \mathbb{R}_+ \times C_0(k)$. Since $\mathbb{R}_+$ is divisible, we have

$$\mathrm{Gal}\,(K/k) \simeq \widehat{A(k)} \simeq C_k/D(k).$$

8.8 The Principal Ideal Theorem

The following result is called the *group theoretical principal ideal theorem*:

Theorem 8.5. *Let G be a finite group. The transfer map $G \longrightarrow [G,G]$ is the trivial map.*

We apply this result to extensions of global fields. Assume first that k is an number field of finite degree, and let K be the maximal abelian extension of k unramified everywhere. A prime $\mathfrak{p}$ of k is said to be completely split in a finite extension k'/k if the number of extensions of $\mathfrak{p}$ to k' equals the degree $(k' : k)$. Let E denote the group of all ideles whose components at the finite primes are units. We now define a homomorphism $J_K \longrightarrow D(K)$ with kernel E by

$$\prod_{\mathfrak{p}} \longmapsto \prod_{\mathfrak{p}\ \text{finite}} \mathfrak{p}^{v_{\mathfrak{p}}(\alpha_{\mathfrak{p}})},$$

where $v_{\mathfrak{p}}$ is the additive valuation with value group $\mathbb{Z}$ and where the product on the right hand side is extended over all finite primes of k. This map induces a homomorphism $C_k \longrightarrow \mathrm{Cl}\,(k)$ with kernel $Ek^\times/k^\times$.

Theorem 8.6. *The extension K/k is finite, and it corresponds by class field theory to the subgroup $Ek^\times/k^\times$ of C_k. Moreover, $\mathrm{Gal}\,(K/k) \simeq \mathrm{Cl}\,(k)$, and a prime $\mathfrak{p}$ in k splits completely in K if and only if $\mathfrak{p}$ is principal.*

Proof. Since $\mathrm{Cl}\,(k)$ is finite, $Ek^\times/k^\times$ has finite index in C_k. According to Section 8.6, the extension of k corresponding to $Ek^\times/k^\times$ is unramified, and conversely to every unramified abelian extension of k there corresponds a subgroup of C_k containing $Ek^\times/k^\times$. This proves the first part of Theorem 8.6.

Let π be a uniformizer of $k_{\mathfrak{p}}$. According to Section 8.6, $\mathfrak{p}$ splits completely if and only if the idele class of π is in $Ek^\times/k^\times$, that is, if and only if $\mathfrak{p}$ is principal. □

We now formulate the p-version of Theorem 8.6. Let $K(p)$ denote the maximal unramified abelian p-extension of k, and $\mathrm{Cl}\,(k)[p']$ the subgroup of $\mathrm{Cl}\,(k)$ consisting of elements whose orders are coprime to p.

Theorem 8.7. $\mathrm{Gal}(K(p)/k)$ *is isomorphic to* $\mathrm{Cl}(k)/\mathrm{Cl}(k)[p']$. *A prime* $\mathfrak{p}$ *in* k *splits completely in* $K(p)$ *if and only if there is an integer* a *coprime to* p *such that* $\mathfrak{p}^a$ *is principal.*

The following theorem is the p-version of the principal ideal theorem:

Theorem 8.8. *If* $\mathfrak{a}$ *is an ideal in* k, *then there exists an integer* a *coprime to* p *such that* $\mathfrak{a}^a$ *is principal in* $K(p)$.

Proof. Let K' denote the maximal unramified abelian p-extension of $K(p)$. Then $\mathrm{Gal}(K(p)/k)$ is the maximal abelian quotient of $\mathrm{Gal}(K'/k)$, and therefore $\mathrm{Gal}(K'/K(p))$ is the commutator subgroup of $\mathrm{Gal}(K'/k)$. By Theorem 8.5, the transfer $\mathrm{Gal}(K'/k) \longrightarrow \mathrm{Gal}(K'/K(p))$ is trivial. By Section 8.6.(ii) this means that every idele class in C_k lies in the subgroup of $C_{K(p)}$ corresponding to $\mathrm{Gal}(K'/K(p))$. By Theorem 8.7, this implies the claim. □

Now let k be a global field of finite type and of characteristic p, and let S be a finite nonempty set of primes of k. Then K^S denotes the maximal abelian unramified extension of k in which all primes in S split completely, and $K^S(p)$ denotes the maximal p-extension of k contained in K^S/k.

Theorem 8.9. K^S/k *is a finite extension corresponding by class field theory to the subgroup*

$$Ek^\times \prod_{\mathfrak{p}\in S} k_\mathfrak{p}^\times / k^\times$$

of C_k, *and* $\mathrm{Gal}(K^S/k)$ *is isomorphic to the quotient group* $\mathrm{Cl}_S(k)$ *of* $\mathrm{Cl}(k)$ *modulo the subgroup generated by primes in* S. *A prime* $\mathfrak{p}$ *in* k *splits completely in* K^S *if and only if there exists an ideal* $\mathfrak{a} \in \langle S\rangle$ *such that* $\mathfrak{p}\mathfrak{a}$ *is principal.*

Proof. This proof is completely analogous to that of Theorem 8.6. □

We now formulate the p-version of Theorem 8.9; to this end, let $\langle S\rangle$ denote the set of ideals in k divisible only by primes in S.

Theorem 8.10. $\mathrm{Gal}(K^S(p)/k)$ *is isomorphic to* $\mathrm{Cl}_S(k)/\mathrm{Cl}_S(k)(p')$. *A prime* $\mathfrak{p}$ *in* k *splits completely in* $K(p)$ *if and only if there is an integer* a *coprime to* p *and an ideal* $\mathfrak{a} \in \langle S\rangle$ *such that* $\mathfrak{p}^a\mathfrak{a}$ *is principal.*

As an analog of Theorem 8.8 we now get

Theorem 8.11. *If* $\mathfrak{a}$ *is an ideal in* k, *then there exists an integer* a *coprime to* p *and an ideal* $\mathfrak{b} \in \langle S\rangle$ *such that* $\mathfrak{a}^a\mathfrak{b}$ *is principal in* $K^S(p)$.

8.9 Cohomology of the Formation Module

Let k be a local or global field of finite type, K/k an arbitrary normal extension and $G = \mathrm{Gal}(K/k)$. In this section we formulate theorems on the structure of $\mathrm{H}^n(G, A(K))$. The transition to infinite extensions follows in general without problems from Theorem 3.16, since $A(K)$ is the inductive limit of the system $\{[\mathrm{Gal}(K_\nu/k), A(K_\nu)]\}$, where the K_ν/k run through all finite normal subextensions of K/k.

The first and the third cohomology group of $A(K)$ vanish. There is a canonical monomorphism $\mathrm{inv}_k : \mathrm{H}^2(G, A(K)) \hookrightarrow \mathbb{Q}/\mathbb{Z}$. *The image of $\alpha \in \mathrm{H}^2(G, A(K))$ under* inv_k *is called the invariant of α. For finite extensions K/k, $\mathrm{H}^2(G, A(K))$ is cyclic of order $(K : k)$.*

There is a connection between the map inv_k and the Artin symbol $(\,\cdot\,, K/k)$: Let K_0/k be the maximal abelian subextension of K/k, $\chi \in \mathrm{H}^1(G, \mathbb{Q}/\mathbb{Z})$ a character of G and $\Delta : \mathrm{H}^1(G, \mathbb{Q}/\mathbb{Z}) \longrightarrow \mathrm{H}^2(G, \mathbb{Z})$ the connecting homomorphism from the cohomology sequence corresponding to the exact sequence

$$0 \longrightarrow \mathbb{Z} \longrightarrow \mathbb{Q} \longrightarrow \mathbb{Q}/\mathbb{Z} \longrightarrow 0.$$

Then Δ is an isomorphism, since $\mathrm{H}^\nu(G, \mathbb{Q}) = 0$ for all $\nu \geq 1$. Now consider the cup product corresponding to the pairing

$$A(K) \times \mathbb{Z} \longrightarrow A(K).$$

Since $A(k) = \mathrm{H}^0(G, A(K))$, we have $a \cup \Delta\chi \in \mathrm{H}^2(G, A(K))$ for $a \in A(k)$. Moreover,

$$\chi(a, K_0/k) = \mathrm{inv}_k(a \cup \Delta\chi). \tag{8.7}$$

Because of (8.7), it is possible to derive the properties of inv_k from those of the Artin symbol and vice versa.

For inv_k, there are the following transformation rules:

(i) If k'/k is an arbitrary algebraic extension, then the diagram

$$\begin{array}{ccc} \mathrm{H}^2(\mathrm{Gal}(K/k), A(K)) & \xrightarrow{\mathrm{inv}_k} & \mathbb{Q}/\mathbb{Z} \\ \downarrow & & \downarrow{\scriptstyle (k':k)} \\ \mathrm{H}^2(\mathrm{Gal}(Kk'/k'), A(Kk')) & \xrightarrow[\mathrm{inv}_{k'}]{} & \mathbb{Q}/\mathbb{Z} \end{array}$$

commutes.

(ii) If k'/k is a finite subextension of K/k, then the corestriction

$$\mathrm{H}^2(\mathrm{Gal}(K/k'), A(K)) \longrightarrow \mathrm{H}^2(\mathrm{Gal}(K/k), A(K))$$

is injective, and the diagram

$$\begin{array}{ccc} \mathrm{H}^2(\mathrm{Gal}\,(K/k'), A(K)) & \longrightarrow & \mathrm{H}^2(\mathrm{Gal}\,(K/k), A(K)) \\ & {\scriptstyle \mathrm{inv}_{k'}} \searrow \quad \swarrow {\scriptstyle \mathrm{inv}_k} & \\ & \mathbb{Q}/\mathbb{Z} & \end{array}$$

commutes.

(iii) If K'/k is a normal subextension of K/k, then the inflation

$$\mathrm{H}^2(\mathrm{Gal}\,(K'/k), A(K')) \longrightarrow \mathrm{H}^2(\mathrm{Gal}\,(K/k), A(K))$$

is injective, and the diagram

$$\begin{array}{ccc} \mathrm{H}^2(\mathrm{Gal}\,(K'/k), A(K')) & \longrightarrow & \mathrm{H}^2(\mathrm{Gal}\,(K/k), A(K)) \\ & {\scriptstyle \mathrm{inv}_{k}} \searrow \quad \swarrow {\scriptstyle \mathrm{inv}_k} & \\ & \mathbb{Q}/\mathbb{Z} & \end{array}$$

commutes.

The first claim of (iii) follows from Theorem 3.14 since $\mathrm{H}^1(G, A(K)) = 0$.

(iv) If $\sigma : K \longrightarrow K^\sigma$ is an isomorphism and σ^* is the induced isomorphism

$$\mathrm{H}^2(\mathrm{Gal}\,(K/k), A(K)) \longrightarrow \mathrm{H}^2(\mathrm{Gal}\,(K^\sigma, k^\sigma), A(K^\sigma)),$$

then the diagram

$$\begin{array}{ccc} \mathrm{H}^2(\mathrm{Gal}\,(K/k), A(K)) & \xrightarrow{\sigma^*} & \mathrm{H}^2(\mathrm{Gal}\,(K^\sigma/k^\sigma), A(K^\sigma)) \\ & {\scriptstyle \mathrm{inv}_{k}} \searrow \quad \swarrow {\scriptstyle \mathrm{inv}_{k^\sigma}} & \\ & \mathbb{Q}/\mathbb{Z} & \end{array}$$

commutes.

Now let k be a local field, K a finite normal unramified extension, $G = \mathrm{Gal}\,(K/k)$, and E the unit group of K. Then $\mathrm{H}^\nu(G, E) = 0$ for all $\nu \geq 1$.

The cohomology sequence corresponding to the exact sequence

$$1 \longrightarrow E \longrightarrow K^\times \xrightarrow{\ v\ } \mathbb{Z} \longrightarrow 0$$

yields an isomorphism

$$\mathrm{H}^2(G, K^\times) \longrightarrow \mathrm{H}^2(G, \mathbb{Z})$$

that we compose with the homomorphism

$$\Delta^{-1} : \mathrm{H}^2(G,\mathbb{Z}) \longrightarrow \mathrm{H}^1(G,\mathbb{Q}/\mathbb{Z}).$$

The group G is cyclic since K/k is unramified, and it is generated by the Frobenius automorphism $\sigma_{K/k}$. We define a monomorphism $\mathrm{H}^1(G,\mathbb{Q}/\mathbb{Z}) \hookrightarrow \mathbb{Q}/\mathbb{Z}$ by

$$\chi \longmapsto \chi(\sigma_{K/k}).$$

By composition of maps we get a map $\varphi : \mathrm{H}^2(G,K^\times) \longrightarrow \mathbb{Q}/\mathbb{Z}$. We claim that $\varphi = \mathrm{inv}_k$. In fact, we have

$$\varphi(\pi \cup \Delta\chi) = \chi(\sigma_{K/k}),$$

and now a comparison with (8.7) proves the claim by Section 8.6.

It can be shown that, in general, inv_k is uniquely determined by these calculations for unramified extensions and the transformation rules (i) and (iii). This constitutes an important step in setting up the cohomology machinery of local class field theory.

8.10 Cohomology of the Multiplicative Group

Now let K/k be a finite extension of global fields and put $G = \mathrm{Gal}\,(K/k)$. For every prime $\mathfrak{p}$ of k fix a prime divisor $\mathfrak{P}$ in K above $\mathfrak{p}$. The embedding $K \hookrightarrow K_{\mathfrak{P}}$ induces a homomorphism

$$\psi_{\mathfrak{p}} : \mathrm{H}^2(G,K^\times) \longrightarrow \mathrm{H}^2(\mathcal{Z}_{\mathfrak{P}},K_{\mathfrak{P}}^\times),$$

where $\mathcal{Z}_{\mathfrak{P}} = \mathrm{Gal}\,(K_{\mathfrak{P}}/k_{\mathfrak{p}})$.

For some $\alpha \in \mathrm{H}^2(G,K^\times)$, $\psi_{\mathfrak{p}}\alpha$ vanishes for almost all $\mathfrak{p}$. In fact, K/k is unramified for almost all $\mathfrak{p}$, hence for almost all $\mathfrak{p}$ a cocycle corresponding to α takes values in the unit group $E_{\mathfrak{P}}$ of $K_{\mathfrak{P}}$. For such a $\mathfrak{p}$, $\psi_{\mathfrak{p}}\alpha$ lies in the image of the map

$$\mathrm{H}^2(\mathcal{Z}_{\mathfrak{P}},E_{\mathfrak{P}}) \longrightarrow \mathrm{H}^2(\mathcal{Z}_{\mathfrak{P}},K_{\mathfrak{P}}^\times),$$

and $\mathrm{H}^2(\mathcal{Z}_{\mathfrak{P}},E_{\mathfrak{P}})$ vanishes for unramified extensions $K_{\mathfrak{P}}/k_{\mathfrak{p}}$. The map

$$\mathrm{H}^2(G,K^\times) \longrightarrow \coprod_{\mathfrak{p}} \mathrm{H}^2(\mathcal{Z}_{\mathfrak{P}},K_{\mathfrak{P}}^\times), \tag{8.8}$$

where the direct sum is over all primes of k and where $\mathfrak{P}$ is the fixed prime divisor of $\mathfrak{p}$, is therefore well defined. Now $\mathrm{inv}_{k_{\mathfrak{p}}} : \mathrm{H}^2(\mathcal{Z}_{\mathfrak{P}},K_{\mathfrak{P}}^\times) \longrightarrow \mathbb{Q}/\mathbb{Z}$ induces a map

$$\coprod_{\mathfrak{p}} \mathrm{H}^2(\mathcal{Z}_{\mathfrak{P}},K_{\mathfrak{P}}^\times) \longrightarrow \mathbb{Q}/\mathbb{Z} \tag{8.9}$$

called the *sum of invariants*.

Hasse's Local-Global Principle states that the sequence

$$0 \longrightarrow \mathrm{H}^2(G, K^\times) \longrightarrow \coprod_{\mathfrak{p}} \mathrm{H}^2(\mathcal{Z}_{\mathfrak{P}}, K_{\mathfrak{P}}^\times) \longrightarrow \mathbb{Q}/\mathbb{Z} \qquad (8.10)$$

formed by composing (8.8) and (8.9) is exact. Transferring this to infinite extensions using Theorem 3.16 poses no problems.

For finite normal extensions K/k the group $\mathrm{H}^3(G, K^\times)$ is cyclic of order $(K:k)/l(K/k)$, where $l(K/k)$ is the least common multiple of the local degrees $(K_{\mathfrak{P}} : k_{\mathfrak{p}})$ for all primes $\mathfrak{P} \mid \mathfrak{p}$. Let L/k be a normal subextension of K/k. The diagram

$$\begin{array}{ccccccccc} 1 & \longrightarrow & L^\times & \longrightarrow & J_L & \longrightarrow & C_L & \longrightarrow & 1 \\ & & \downarrow & & \downarrow & & \downarrow & & \\ 1 & \longrightarrow & K^\times & \longrightarrow & J_K & \longrightarrow & C_K & \longrightarrow & 1 \end{array}$$

induces the commutative diagram

$$\begin{array}{ccc} \mathrm{H}^2(\mathrm{Gal}\,(L/k), C_L) & \longrightarrow & \mathrm{H}^3(\mathrm{Gal}\,(L/k), L^\times) \\ {\scriptstyle\mathrm{inf}}\downarrow & & \downarrow{\scriptstyle\mathrm{inf}} \\ \mathrm{H}^2(\mathrm{Gal}\,(K/k), C_K) & \longrightarrow & \mathrm{H}^3(\mathrm{Gal}\,(K/k), K^\times) \end{array}$$

Here the horizontal maps are epimorphisms. The inflation in the left column is a monomorphism of cyclic groups. The image of the inflation in the right column thus vanishes if $(K:k)/l(K/k)$ divides $(K:L)$, i.e., if $(L:k) \mid l(K/k)$.

8.11 Norm Residue Symbol

Let k be a local field; we now have a closer look at its Artin symbol, which is also called the *norm residue symbol* in this case.

Assume that k contains the n-th roots of unity, where n is coprime to the characteristic of k, and let ζ_n denote a primitive n-th root of unity. Moreover, let $K = k(\sqrt[n]{\alpha})$, $\alpha \in k$, be a cyclic extension of k. Then the norm residue symbol (α, β) for n-th powers is defined for $\alpha, \beta \in k^\times$ by

$$(\beta, K/k)\,\sqrt[n]{\alpha} = (\alpha, \beta)\,\sqrt[n]{\alpha}.$$

Then (α, β) is an n-th root of unity with the following properties:

(i) $(\alpha\alpha', \beta) = (\alpha, \beta)(\alpha', \beta)$;

(ii) $(\alpha, \beta\beta') = (\alpha, \beta)(\alpha, \beta')$;

(iii) $(\alpha, \beta)(\beta, \alpha) = 1$;

(iv) If $(\alpha, \beta) = 1$ for all $\beta \in k^\times$, then $\alpha \in k^{\times n}$;

(v) We have $(\alpha, \beta) = 1$ if and only if β is a norm in $k(\sqrt[n]{\alpha})/k$.

The properties (ii), (iv) and (v) follow immediately from the properties of the Artin symbol mentioned in Section 8.6. We shall prove (iii) below, and then (i) follows from (ii) and (iii).

Next we realize (α, β) as a cup product of elements of a 1-dimensional cohomology group. To this end we restrict ourselves to the case where $n = q$ is a power of p. Let G be the Galois group of the maximal p-extension $\widehat{k}$ of k.

To each $\alpha \in k^\times$ we associate an element $\chi_\alpha \in \mathrm{H}^1(G, \mathbb{Z}/q)$ by

$$g(\sqrt[q]{\alpha}) = \zeta_q^{\chi_\alpha(g)} \sqrt[q]{\alpha}, \quad g \in G.$$

This defines a homomorphism $\varphi : k^\times \longrightarrow \mathrm{H}^1(G, \mathbb{Z}/q)$ with kernel $k^{\times q}$.

Now we prove that the group μ_q of q-th roots of unity in k is isomorphic to $\mathrm{H}^2(G, \mathbb{Z}/q)$, and to this end we consider the exact "Kummer sequence"

$$0 \longrightarrow \mathbb{Z}/q \xrightarrow{\lambda} \widehat{k}^\times \xrightarrow{q} \widehat{k}^\times \longrightarrow 1, \tag{8.11}$$

where $\widehat{k}$ denotes the maximal separable p-extension of k (see Chap. 9), λ is defined by $\lambda(a) = \zeta_n^a$, and the map q stands for raising to the q-th power. From (8.11) we get the cohomology sequence

$$0 \longrightarrow \mathrm{H}^1(G, \widehat{k}^\times) \longrightarrow \mathrm{H}^2(G, \mathbb{Z}/q) \xrightarrow{\lambda^*} \mathrm{H}^2(G, \widehat{k}^\times) \xrightarrow{q} \mathrm{H}^2(G, \widehat{k}^\times).$$

Since, by Section 8.9, the image $\mathrm{H}^2(G, \widehat{k}^\times)[q]$ of λ^* is cyclic of order q, the desired result follows. According to Section 8.9,

$$\varepsilon \longmapsto \zeta_q^{(q \operatorname{inv}_k \varepsilon)}, \quad \varepsilon \in \mathrm{H}^2(G, \widehat{k}^\times),$$

defines an isomorphism $\iota : \mathrm{H}^2(G, \widehat{k}^\times) \longrightarrow \mu_q$. Therefore the composition $\psi = \iota \circ \lambda^* : \mathrm{H}^2(G, \mathbb{Z}/q) \longrightarrow \mu_q$ is an isomorphism.

We now can prove the interpretation of the norm residue symbol we were looking for:

Theorem 8.12. *For all $\alpha, \beta \in k^\times$, we have*

$$(\alpha, \beta) = \psi(\varphi\alpha \cup \varphi\beta)^{-1}.$$

Proof. By definition we have

$$(\alpha, \beta) = \zeta_q^A, \quad = \chi'_\alpha(\beta, k(\sqrt[q]{\alpha})/k),$$

where $\chi'_\alpha \in \mathrm{H}^1(\mathrm{Gal}\,(k(\sqrt[q]{\alpha})/k, \mathbb{Z}/q)$ is induced by χ_α. By (8.7) and the transformation rule (iii) in Section 8.9 we have

$$(\alpha,\beta)=\zeta_q^{(q\,\mathrm{inv}_k(\beta\cup\Delta\chi'_\alpha))}=\iota(\beta\cup\Delta\chi_\alpha).$$

The map $\Delta\chi_\alpha$ is represented by the cocycle

$$g,h\longmapsto\frac{1}{q}\Big(\widetilde{\chi_\alpha(g)}+\widetilde{\chi_\alpha(h)}-\widetilde{\chi_\alpha(gh)}\Big),$$

where $\tilde{z}$ denotes a representative of $z\in\mathbb{Z}/q$ in $\mathbb{Z}$.

For the proof of Theorem 8.12 it therefore remains to prove that

$$\beta\cup\Delta\chi_\alpha=-\lambda^*(\chi_\alpha\cup\chi_\beta).$$

Now $\lambda^*(\chi_\alpha\cup\chi_\beta)$ is represented by the cocycle

$$g,h\longmapsto\zeta_q^{\chi_\alpha(g)\chi_\beta(h)}\;=\;h(\sqrt[q]{\beta})^{\widetilde{\chi_\alpha(g)}}/\sqrt[q]{\beta}^{\widetilde{\chi_\alpha(g)}},\tag{8.12}$$

and $\beta\cup\Delta\chi_\alpha$ is represented by the cocycle

$$g,h\longmapsto\beta^{A/q}=\sqrt[q]{\beta}^{A},\quad A=\widetilde{\chi_\alpha(g)}+\widetilde{\chi_\alpha(h)}-\widetilde{\chi_\alpha(gh)}.\tag{8.13}$$

The product of (8.12) and (8.13) is a coboundary, and this implies the theorem. □

The anti-commutativity of the cup product implies by Theorem 8.12 property (iii) of the norm residue symbol.

As a supplement to theorem 8.12 we prove the following

Theorem 8.13. *If $B:\mathrm{H}^1(G,\mathbb{Z}/q)\longrightarrow\mathrm{H}^2(G,\mathbb{Z}/q)$ is the map induced by the exact sequence*

$$0\longrightarrow\mathbb{Z}/q\xrightarrow{\;q\;}\mathbb{Z}/q^2\longrightarrow\mathbb{Z}/q\longrightarrow 0,$$

then

$$(\alpha,\zeta_q)=\psi(B\varphi\alpha).$$

Proof. As in the proof of Theorem 8.12, we have to show that

$$\zeta_q\cup\Delta\chi_\alpha=\lambda^*(B\varphi\alpha).$$

Now $B\varphi\alpha$ is represented by the cocycle

$$g,h\longmapsto\frac{1}{q}\Big(\widetilde{\chi_\alpha(g)}+\widetilde{\chi_\alpha(h)}-\widetilde{\chi_\alpha(gh)}\Big)+q\mathbb{Z}.$$

This implies the claim. □

9. The Maximal p-Extension

The maximal p-extension $\widehat{k}$ of a field k is the compositum (inside a fixed algebraic closure of k) of all finite p-extensions of k, i.e., of all normal (separable) extensions of k with p-power degree.

The extension $\widehat{k}$ is closed with respect to p-extensions. In fact, if K is a p-extension of $\widehat{k}$, then so are its conjugates over k, and therefore the normal closure N of K/k is a p-extension: thus $N = K = \widehat{k}$.

In the following, we denote $\mathrm{Gal}(\widehat{k}/k)$ by G_k. The main dichotomy is defined by the question whether k contains the p-th roots of unity μ_p or not. For this reason we introduce the function

$$\delta(k) = \begin{cases} 1 & \text{if } k \text{ contains } \mu_p, \\ 0 & \text{otherwise.} \end{cases}$$

Let $\chi(k)$ denote the characteristic of k. For fields k with $\chi(k) = p$ we put $\delta(k) = 0$. In this chapter, we are interested mainly in the cases where G_k is free.

9.1 Fields of Characteristic p

We want to prove the following

Theorem 9.1. *If k is a field of characteristic p, then G_k is a free pro-p group with generator rank*

$$\dim_{\mathbb{F}_p} k^+/\wp(k^+), \tag{9.1}$$

where we have put

$$\wp(x) = x^p - x. \tag{9.2}$$

Proof. Consider the "Artin-Schreier sequence"

$$0 \longrightarrow \mathbb{F}_p \longrightarrow \widehat{k}^+ \xrightarrow{\ \wp\ } \widehat{k}^+ \longrightarrow 0.$$

This sequence is exact. This is obvious except for the surjectivity of $\wp$, which follows from the fact that

$$x^p - x = \alpha, \quad \alpha \in \widehat{k}^+, \tag{9.3}$$

has a solution $\xi \in \widehat{k}$: If ξ is a solution, then so is $\xi + a$ for any $a \in \mathbb{F}_p$; this implies that $\widehat{k}(\xi)/\widehat{k}$ is a p-extension, hence $\xi \in \widehat{k}$. The corresponding cohomology sequence gives

$$k^+ \xrightarrow{\wp} k^+ \longrightarrow \mathrm{H}^1(G_k) \longrightarrow 0$$

because of Theorem 3.17, and this implies (9.1). The cohomology sequence also shows

$$\mathrm{H}^2(G_k) = 0,$$

and this implies by Theorem 4.12 that G_k is free. □

9.2 Fields Containing the p-th Roots of Unity

Now assume that char $k \neq p$, and that k contains the group μ_p of p-th roots of unity. Let ζ_p denote a primitive p-th root of unity.

Theorem 9.2. *We have* $\mathrm{H}^2(G_k) \simeq \mathrm{H}^2(G_k, \widehat{k}^\times)[p]$, *and the generator rank of* G_k *equals*

$$\dim_{\mathbb{F}_p} k^\times / k^{\times p}. \tag{9.4}$$

Proof. Consider the "Kummer sequence"

$$0 \longrightarrow \mathbb{F}_p \xrightarrow{\varphi} \widehat{k}^\times \xrightarrow{p} \widehat{k}^\times \longrightarrow 1, \tag{9.5}$$

where the cyclotomic character φ is defined by

$$\varphi(a) = \zeta_p^a, \quad a \in \mathbb{F}_p$$

and where p stands for raising to the p-th power.

The sequence (9.5) is exact: this is obvious except for the surjectivity of p, which follows from the fact that the equation

$$x^p = \alpha$$

has a solution $\xi \in \widehat{k}$ for all $\alpha \in \widehat{k}^\times$: the extension $\widehat{k}(\xi)/\widehat{k}$ is a p-extension, hence $\xi \in \widehat{k}$.

The cohomology sequence corresponding to (9.5) implies by Theorem 3.18 that

$$k^\times \xrightarrow{p} k^\times \longrightarrow \mathrm{H}^1(G_k) \longrightarrow 0$$

is exact, and this in turn implies (9.4). Similarly,

$$0 \longrightarrow \mathrm{H}^2(G_k) \longrightarrow \mathrm{H}^2(G_k, \widehat{k}^\times) \xrightarrow{p} \mathrm{H}^2(G_k, \widehat{k}^\times),$$

where p denotes the map multiplication by p, implies the first claim of Theorem 9.2. □

Next we look at the cases where $H^2(G_k, \widehat{k}^\times)[p]$ vanishes. This happens in particular if the base field k is "sufficiently large". A local field k is called *p-infinite* if it is the union of finite local fields k_i, $i \in I$, where for any $i \in I$ there exists a $j \in I$ such that $p \mid (k_j : k_i)$.

Theorem 9.3. *If k is a p-infinite local field, then G_k is a free pro-p group.*

Proof. Because of Theorem 9.2 and Theorem 4.12 it suffices to prove that $H^2(G_k, \widehat{k}^\times)[p] = 0$.

The cohomology group $H^2(G_k, \widehat{k}^\times)$ is isomorphic to $\varinjlim H^2(G_k, k_i^\times)$. It is therefore sufficient to show that $\varinjlim H^2(G_k, k_i^\times)[p]$ vanishes. But if α is an element in $H^2(G_{k_i}, k_i^\times)[p]$ and $p \mid (k_j : k_i)$, then by 8.9, (I), α is mapped to 0 under

$$H^2(\mathrm{Gal}(\widehat{k}_i/k_i), \widehat{k}_i^\times) \longrightarrow H^2(\mathrm{Gal}(\widehat{k}_i k_j/k_j), (\widehat{k}_i k_j)^\times),$$

so in particular it is mapped to 0 under

$$H^2(\mathrm{Gal}(\widehat{k}_i/k_i), \widehat{k}_i^\times) \longrightarrow H^2(\mathrm{Gal}(\widehat{k}_j/k_j), \widehat{k}_j^\times).$$

This implies our claim. □

Theorem 9.4. *Let k be a global field with the property that $k_\mathfrak{p}$ is p-infinite for all finite places $\mathfrak{p}$ of k. If k is an algebraic number field and $p = 2$, then assume that k is totally complex, i.e., that $k_\mathfrak{p} = \mathbb{C}$ for all infinite places $\mathfrak{p}$ of k. Then G_k is a free pro-p group.*

Proof. $H^2(G_k, k^\times)$ is isomorphic to $\varinjlim H^2(G_{k_i}, k_i^\times)$, where k_i runs through all global subfields of k of finite type. Thus it is sufficient to prove that $\varinjlim H^2(G_{k_i}, k_i^\times)[p]$ vanishes.

According to (8.10), there is a commutative diagram

$$\begin{array}{ccccc} 0 & \longrightarrow & H^2(G_{k_i}, \widehat{k}_i^\times) & \longrightarrow & \coprod_{\mathfrak{p}_i} H^2(Z_{\mathfrak{P}_i}, \widehat{k}_{i\,\mathfrak{P}_i}^\times) \\ & & \downarrow & & \downarrow \varphi \\ 0 & \longrightarrow & H^2(G_{k_j}, \widehat{k}_j^\times) & \longrightarrow & \coprod_{\mathfrak{p}_j} H^2(Z_{\mathfrak{P}_j}, \widehat{k}_{j\,\mathfrak{P}_j}^\times) \end{array} \tag{9.6}$$

for $k_i \subseteq k_j$, where $\mathfrak{p}_i$ and $\mathfrak{p}_j$ run through all places of k_i and k_j, respectively, and where $\mathfrak{P}_i$ and $\mathfrak{P}_j$ are fixed primes above $\mathfrak{p}_i$ and $\mathfrak{p}_j$. The map φ is induced by the natural homomorphism

$$\varphi_{\mathfrak{p}_j|\mathfrak{p}_i} : H^2(Z_{\mathfrak{P}_i}, \widehat{k}_{i\,\mathfrak{P}_i}^\times) \longrightarrow H^2(Z_{\mathfrak{P}_j}, \widehat{k}_{j\,\mathfrak{P}_j}^\times)$$

for $\mathfrak{p}_j \mid \mathfrak{p}_i$, i.e., we put

$$\varphi\Big(\sum_{\mathfrak{p}_i} \alpha_{\mathfrak{p}_i}\Big) = \sum_{\mathfrak{p}_i} \sum_{\mathfrak{p}_j|\mathfrak{p}_i} \varphi_{\mathfrak{p}_j|\mathfrak{p}_i}(\alpha_{\mathfrak{p}_i})$$

for $\alpha_{\mathfrak{p}_i} \in \mathrm{H}^2(Z_{\mathfrak{P}_i}, \widehat{k}^\times_{i\,\mathfrak{P}_i})$. The claim now follows from (9.6) similarly as in the proof of Theorem 9.3. □

Example 9.5. The assumptions of Theorems 9.3 and 9.4 are satisfied if k contains μ_{p^∞}, the group of all roots of unity of p-power order.

9.3 Fields not Containing the p-th Roots of Unity

Now assume that the base field k has characteristic different from p and that it does not contain the p-th roots of unity μ_p. We put $k' = k(\zeta_p)$, and we have the following reduction theorem:

Theorem 9.6. *Assume that the restriction*

$$\mathrm{res}\ : \mathrm{H}^2(\mathrm{Gal}\,(\widehat{k'}/k) \longrightarrow \mathrm{H}^2(\mathrm{Gal}\,(\widehat{k'}/\widehat{k}k'))$$

is trivial. Then $\mathrm{H}^2(G_k) \simeq \mathrm{H}^2(G_{k'})^{\mathrm{Gal}\,(k'/k)}$.

Proof. We first show that $\mathrm{H}^2(\mathrm{Gal}\,(\widehat{k'}/\widehat{k}))$ vanishes. To this end we consider a $\lambda \in \mathrm{H}^1(\mathrm{Gal}\,(\widehat{k'}/\widehat{k}))$, i.e., a homomorphism $\lambda : \mathrm{Gal}\,(\widehat{k'}/\widehat{k}) \longrightarrow \mathbb{F}_p$. Then $\ker\lambda$ has index 1 or p in $\mathrm{Gal}\,(\widehat{k'}/\widehat{k})$ and is a normal subgroup of $\mathrm{Gal}\,(\widehat{k'}/k)$; thus the fixed field of $\ker\lambda$ is a p-extension of $\widehat{k}$, hence equal to $\widehat{k}$. This implies that $\ker\lambda = \mathrm{Gal}\,(\widehat{k'}/\widehat{k})$, i.e., $\lambda = 0$.

Next we look at the inflation

$$\mathrm{inf} : \mathrm{H}^2(\mathrm{Gal}\,(\widehat{k}/k)) \longrightarrow \mathrm{H}^2(\mathrm{Gal}\,(\widehat{k'}/k)). \tag{9.7}$$

Since $\mathrm{H}^1(\mathrm{Gal}\,(\widehat{k'}/\widehat{k})) = 0$, Theorem 3.14 gives us an exact sequence

$$0 \longrightarrow \mathrm{H}^2(\mathrm{Gal}\,(\widehat{k}/k)) \xrightarrow{\ \mathrm{inf}\ } \mathrm{H}^2(\mathrm{Gal}\,(\widehat{k'}/k)) \xrightarrow{\ \mathrm{res}\ } \mathrm{H}^2(\mathrm{Gal}\,(\widehat{k'}/\widehat{k})).$$

On the other hand, the restriction

$$\mathrm{res}\ : \mathrm{H}^2(\mathrm{Gal}\,(\widehat{k'}/\widehat{k})) \longrightarrow \mathrm{H}^2(\mathrm{Gal}\,(\widehat{k'}/\widehat{k}k'))$$

is injective by Theorem 3.15 since the degree $(\widehat{k}k' : \widehat{k})$ is coprime to p. Thus if

$$\mathrm{H}^2(\mathrm{Gal}\,(\widehat{k'}/k)) \longrightarrow \mathrm{H}^2(\mathrm{Gal}\,(\widehat{k'}/\widehat{k}k'))$$

is trivial, then so is

$$\mathrm{H}^2(\mathrm{Gal}\,(\widehat{k'}/k)) \longrightarrow \mathrm{H}^2(\mathrm{Gal}\,(\widehat{k'}/\widehat{k})),$$

and therefore (9.7) is an isomorphism.

Moreover, $p \nmid (k' : k)$ implies, again by Theorem 3.17, that the restriction

$$\mathrm{res}\ : \mathrm{H}^2(\mathrm{Gal}\,(\widehat{k'}/k)) \longrightarrow \mathrm{H}^2(\mathrm{Gal}\,(\widehat{k'}/k')) \tag{9.8}$$

is an isomorphism.

The desired isomorphism $\mathrm{H}^2(G_k) \longrightarrow \mathrm{H}^2(G_{k'})^{\mathrm{Gal}\,(k'/k)}$ now follows upon composing (9.7) and (9.8). □

According to Theorems 9.3 and 9.4, we have $\mathrm{H}^2(\mathrm{Gal}(\widehat{k'}/\widehat{k}k')) = 0$ if k is a local or a global field that is totally complex if $p = 2$. The assumptions of Theorem 9.6 are therefore satisfied. In particular, we have

Theorem 9.7. *Let k be a p-infinite local field or a global field with the property that $k_{\mathfrak{p}}$ is p-infinite for all finite places $\mathfrak{p}$ of k. Then G_k is a free pro-p group.*

Example 9.8. The assumptions of Theorem 9.7 are satisfied if k contains the field of p^n-th roots of unity for all $n \geq 2$.

10. Local Fields of Finite Type

In this chapter we study the maximal p-extension more closely in the case where k is a local field of finite type. Let $\mathfrak{p}$ denote the prime ideal of k and $\chi(\mathfrak{p})$ the characteristic of the residue class field of k.

The maximal unramified p-extension T_k/k is a subextension of $\widehat{k}/k$. By Sections 8.3 and 8.6, the Galois group $\mathrm{Gal}\,(T_k/k)$ is isomorphic to $\mathbb{Z}_p$ with a distinguished generator σ_k, the Frobenius automorphism of T_k/k. Thus the problem arises of giving a group-theoretical description of the pro-p group $G_k = \mathrm{Gal}\,(\widehat{k}/k)$ with the distinguished normal subgroup $\mathcal{T}_k = \mathrm{Gal}\,(\widehat{k}/T_k)$, where the quotient group $G_k/\mathcal{T}_k$ is canonically isomorphic to $\mathbb{Z}_p$.

In the most difficult case where k is a $\mathfrak{p}$-adic number field containing the p-th roots of unity and for which $\chi(\mathfrak{p}) = p$, we shall solve this problem only partially.

10.1 The Case $\chi(\mathfrak{p}) \neq p$

We first consider the case where the characteristic of the residue class field does not equal p.

If k does not contain the p-th roots of unity, i.e., if

$$N(\mathfrak{p}) \not\equiv 1 \pmod{\mathfrak{p}},$$

then $\widehat{k}/k$ is unramified by Section 8.5.

Theorem 10.1. *Assume that $\chi(\mathfrak{p}) \neq p$ and $\delta(k) = 0$. Then $G_k \simeq \mathbb{Z}_p$ and $\mathcal{T}_k = 1$.*

Now we assume that k contains the p-th roots of unity; then

$$N(\mathfrak{p}) \equiv 1 \pmod{\mathfrak{p}}.$$

By adjoining a primitive p^n-th root of unity ζ_{p^n} to k we get an unramified p-extension whose degree tends to ∞ with n, that is, we have

$$T_k = \bigcup_{n=1}^{\infty} k(\zeta_{p^n}).$$

Thus by Section 8.5, every finite subextension of $\widehat{k}/k$ can be generated by two elements ζ_{p^n} and $\sqrt[p^m]{\pi\zeta}$, where π is a fixed prime element of k and ζ a root of unity in T_k. Thus $\sqrt[p^m]{\zeta}$ is also in T_k, hence $\widehat{k}$ is the union of the normal extensions

$$K_n = k(\zeta_{p^n}, \sqrt[p^n]{\pi}), \quad n = 1, 2, \ldots.$$

We normalize the roots of unity in such a way that

$$\zeta_{p^n}^{p^m} = \zeta_{p^{n-m}}$$

for $n \geq m$. Then $\mathrm{Gal}(K_n/k)$ is generated by two automorphisms σ_n, τ_n defined by

$$\sigma_n(\zeta_{p^n}) = \zeta_{p^n}^{N(\mathfrak{p})}, \ \sigma_n(\sqrt[p^n]{\pi}) = \sqrt[p^n]{\pi}$$

and

$$\tau_n(\zeta_{p^n}) = \zeta_{p^n}, \tau_n(\sqrt[p^n]{\pi}) = \zeta_{p^n}\sqrt[p^n]{\pi}.$$

Then σ_n is a lift of the Frobenius automorphism of $k(\zeta_{p^n})/k$ and τ_n is a generator of the cyclic inertia group of K_n/k. The automorphisms σ_n and τ_n satisfy the relations

$$\sigma_n \tau_n \sigma_n^{-1} = \tau_n^{N(\mathfrak{p})}. \tag{10.1}$$

The group G_k is the inverse limit of the system $\{\mathrm{Gal}(K_n/k) \mid n = 1, 2, \ldots\}$. Thus G_k is a pro-p group with generators

$$\sigma = \prod_{n=1}^{\infty} \sigma_n, \quad \tau = \prod_{n=1}^{\infty} \tau_n$$

satisfying the relation

$$\sigma\tau\sigma^{-1} = \tau^{N(\mathfrak{p})}. \tag{10.2}$$

The automorphism σ is a lift of the Frobenius automorphism of the maximal unramified p-extension T_k/k, and τ generates the inertia subgroup $\mathcal{T}_k$ of G_k. By construction, $\mathcal{T}_k \simeq \mathbb{Z}_p$. Every lift σ' of the Frobenius automorphism of T_k/k has the form

$$\sigma' = \sigma\tau^a, \quad a \in \mathbb{Z}_p,$$

and therefore also satisfies (10.2). The relation (10.2) is also conserved upon replacing τ by another generator of $\mathcal{T}_k$. This proves the following theorem:

Theorem 10.2. *If $\chi(\mathfrak{p}) \neq p$ and $\delta(k) = 1$, then G_k is a pro-p group with two generators σ and τ satisfying the relations*

$$\tau^{N(\mathfrak{p})-1}[\tau^{-1}, \sigma^{-1}] = 1.$$

Here σ is an arbitrary lift to $\widehat{k}$ of the Frobenius automorphism of the maximal unramified p-extension of k, and τ is an arbitrary generator of the inertia subgroup of $\widehat{k}/k$, which is isomorphic to $\mathbb{Z}_p$.

By Theorem 10.2, G_k is a pro-p group with generator rank 2 and relation rank 1. This result could also have been derived from Theorem 9.2.

Actually, we have the following theorem:

Theorem 10.3. *If k is a local field of finite type with characteristic $\neq p$ and $\delta(k) = 1$, then G_k is a pro-p group of cohomological dimension 2 with generator rank 2 if $\chi(\mathfrak{p}) \neq p$, and $(k : \mathbb{Q}_p) + 2$ if $\chi(\mathfrak{p}) = p$, and with relation rank 1.*

Proof. The generator rank is computed as in Section 8.5 using Theorem 9.2. By Section 8.9 and Theorem 9.2, we have $r(G_k) = 1$. Moreover,

$$\chi_2(G_k) = \begin{cases} 0 & \text{if } \chi(\mathfrak{p}) \neq p, \\ -(k : \mathbb{Q}_p) & \text{if } \chi(\mathfrak{p}) = p. \end{cases}$$

Now let U be a subgroup of finite index in G_k. Then the field $K(U)$ fixed by U is a finite p-extension of k, and in the second case we have

$$\chi_2(U) = -(K(U) : \mathbb{Q}_p) = -(K(U) : k)(k : \mathbb{Q}_p) = (G_k : U)\chi_2(G_k);$$

in the first case we have $\chi_2(U) = (G_k : U)\chi_2(G_k)$ for trivial reasons. By Theorem 5.5, the cohomological dimension of G_k is therefore equal to 2. □

10.2 The Case $\chi(\mathfrak{p}) = p, \delta(k) = 0$

If the characteristic of k is a prime p, then the characteristic of the residue class field must also be $\chi(\mathfrak{p}) = p$. In this case, we are therefore in the situation of Theorem 9.1. We have

Theorem 10.4. *Let k be a local field of finite type with characteristic p; then G_k is a finitely generated free pro-p group.*

We can choose the system of generators in the form $\{\sigma, \tau_1, \tau_2, \ldots\}$, where σ is a lift of the Frobenius automorphism of T_k/k, and where $\tau_1, \tau_2, \ldots \in \mathrm{Gal}\,(k/T_k)$.

Proof. By Section 8.7, the generator rank $d(G_k)$ of G_k is given by $d(G_k) = \dim \mathrm{H}^1(G_k) = \dim(G_k/G_k^p[G_k, G_k])^* = \dim(k^\times/k^{\times p})^*$. By Section 8.5, $k^\times/k^{\times p}$ is direct product of countably many cyclic groups of order p; this implies that $d(K_k)$ is countable. This can also be proved by using Theorem 9.1 to compute the dimension of $k^+/\wp k^+$.

Let π be a prime element of k and $\varepsilon_1, \varepsilon_2, \ldots$ a system of principal units of k such that every principal unit ε of k can be written uniquely in the form

$$\varepsilon = \prod_{\nu=1}^{\infty} \varepsilon_\nu^{a_\nu}, \quad a_\nu \in \mathbb{Z}_p.$$

Moreover, let $\overline{k}$ denote the maximal abelian p-extension of k. By Section 8.7,

$$\sigma' = (\pi, \overline{k}/k), \;\; \tau'_\nu = (\varepsilon_\nu, \overline{k}/k), \quad \nu = 1, 2, \ldots$$

form a minimal system of generators of $\mathrm{Gal}\,(\overline{k}/k)$, where σ' is a lift of the Frobenius automorphism of T_k/k, and where the τ'_ν are elements of the inertia group of $\overline{k}/k$. Lifting these automorphisms to $\widehat{k}$ via Theorem 4.10 gives the desired system of generators of G_k. □

Now assume that k has characteristic 0, i.e., that k is a finite extension of $\mathbb{Q}_p$, and suppose that $\delta(k) = 0$. We have

Theorem 10.5. *Let k be a $\mathfrak{p}$-adic number field with $\chi(\mathfrak{p}) = p$ not containing the p-th roots of unity. Then G_k is a free pro-p group with generator rank $(k : \mathbb{Q}_p) + 1 = n + 1$.*

The system of generators of G_k can be chosen in the form $\{\sigma, \tau_1, \tau_2, \ldots \tau_n\}$, where σ is a lift of the Frobenius automorphism of T_k/k, and where τ_1, τ_2, $\ldots$, τ_n are elements of the inertia group of $\overline{k}/k$.

Proof. The proof of Theorem 10.5 is, as far as the generators are concerned, completely analogous to the proof of Theorem 10.4. By Section 8.5, we find that the generator rank of G_k is

$$d(G_k) = \dim(k^\times/k^{\times p})^* = (k : \mathbb{Q}_p) + 1. \tag{10.3}$$

In order to prove that G_k is free we have to show that cd $G_k \leq 1$, i.e., $\mathrm{H}^2(G_k) = 0$. We shall give two different proofs.

1. Just as in the proof of Theorem 10.3 we can show that

$$\chi_1(U) = (G_k : U)\chi_1(G_k)$$

for every subgroup U of finite index in G_k. Thus cd $G_k \leq 1$ by Theorem 5.5.

2. We apply Theorem 9.6. The conditions are satisfied by Theorem 9.3. Thus

$$\mathrm{H}^2(G_k) \simeq \mathrm{H}^2(G_{k'})^{\mathrm{Gal}\,(k'/k)}.$$

We now show that $\mathrm{H}^2(G_{k'})^{\mathrm{Gal}\,(k'/k)} = 0$ and start by remarking that, by Section 8.9, (iv), we have

$$\mathrm{H}^2(G_{k'}, \widehat{k}'^\times)^{\mathrm{Gal}\,(k'/k)} = \mathrm{H}^2(G_{k'}, \widehat{k}'^\times). \tag{10.4}$$

On the other hand, Section 9.2 provides us with an isomorphism

$$\varphi^* : \mathrm{H}^2(G_{k'}) \longrightarrow \mathrm{H}^2(G_{k'}, \widehat{k}'^\times)[p]$$

induced by the map

$$\varphi(a) = \zeta_p^a, \quad a \in \mathbb{F}_p.$$

For $s \in \mathrm{Gal}\,(k'/k)$, put

$$s\zeta_p = \zeta_p^{a(s)}, \quad a(s) \in \mathbb{F}_p.$$

Then

$$s\varphi^* = a(s)\varphi^* s. \tag{10.5}$$

In fact, $s\varphi^*$ is induced by the morphism

$$G_{k'} \longleftarrow G'_k : g \longmapsto \overline{s}^{-1} g \overline{s}, g \in G_{k'},$$
$$\mathbb{F}_p \longrightarrow \widehat{k}'^{\times} : \quad a \longmapsto \zeta_p^a, a \in \mathbb{F}_p,$$

where $\overline{s}$ is a lift of s to $\mathrm{Gal}\,(\widehat{k}'/k)$.

For $\alpha \in \mathrm{H}^2(G_{k'})^{\mathrm{Gal}\,(k'/k)}$, (10.4) and (10.5) show that

$$a(s)\varphi^*(\alpha) = \varphi^*(\alpha) = 0,$$

hence $\alpha = 0$. □

10.3 The Case $\chi(\mathfrak{p}) = p, \delta(k) = 1$

We first assume that

$$n = (k : \mathbb{Q}_p) \equiv 0 \pmod 2.$$

For $p \neq 2$, this is always true since $(\mathbb{Q}_p(\zeta_p) : \mathbb{Q}_p) = p - 1$.

Let $q = p^\kappa$ be the highest power of p such that k contains the q-th roots of unity. Below, we only give a description of G_k mod $G_k^{(3,q)} \cap G_k^{(1)}$.

We prove three lemmas.

Lemma 10.6. *Let ζ_q be a primitive q-th root of unity and q' the maximal power of p such that*

$$k(\sqrt[q']{\zeta_q})/k$$

is unramified. Moreover, let g be a rational integer such that

$$\sigma \sqrt[q']{\zeta_q} = \sqrt[q']{\zeta_q}^{\,g}, \tag{10.6}$$

where σ is the Frobenius automorphism of $k(\sqrt[q]{\zeta_q})/k$. Then $g \equiv 1 \pmod q$, and there is a basis $\{\pi, \alpha_0, \alpha_1, \ldots, \alpha_n\}$ of $k^\times/k^{\times q}$ satisfying the following conditions:

(i) *π is a uniformizer of k;*

(ii) *α_0 is q-*primary*, i.e., $k(\sqrt[q]{\alpha_0})/k$ is unramified;*

(iii) *$\alpha_1, \ldots, \alpha_n$ are units of k;*

(iv) *Let $(\cdot\,,\cdot)$ denote the q-th power norm residue symbol in k; then*

$$(\alpha_0, \pi) = \zeta_q,$$
$$(\alpha_{2\nu-1}, \alpha_{2\nu}) = \zeta_q, \nu = 1, 2, \ldots, n/2;$$

for all other pairs of different basis elements the norm residue symbol equals 1;

(v) $\sigma \sqrt[q]{\alpha_0} = \zeta_q \sqrt[q]{\alpha_0}$;

(vi) $\zeta_q = \alpha_0^{(q-1)/n} \alpha_1^{q'}$.

Proof. It follows form (10.6) that

$$\sigma\zeta_q = \zeta_q = \zeta_q^g,$$

hence we have $g \equiv 1 \pmod q$.

Let α_0 be a q-primary element of k satisfying (v); in particular, we have $(k(\sqrt[q]{\alpha_0}) : k) = q$.

We first assume that $q' < q$. Then

$$k(\sqrt[q']{\zeta_q}) = k(\sqrt[q']{\alpha_0}),$$

hence

$$\zeta_q = \alpha_0^\kappa \alpha_1^{q'}, \quad \alpha_1 \in k$$

for some unit α_1. From (v) we get

$$\sigma \sqrt[q']{\zeta_q} = \sqrt[q']{\zeta_q}^{\,g} = \zeta_q^{\kappa q q'^{-1}} \sqrt[q']{\zeta_q}$$

and therefore

$$\frac{g-1}{q'} \equiv \frac{xq}{q'} \bmod q,$$

that is,

$$x \equiv \frac{g-1}{q} \bmod q'.$$

Thus we may assume without loss of generality that $x = \frac{g-1}{q}$.

α_1 does not have the form $\alpha_0^a \beta^p$ for $a \in \mathbb{Z}$, $\beta \in k$, since this would imply that $k(\sqrt[q'p]{\zeta_q})/k$ is unramified in contradiction to the definition of q'.

Now consider the case $q' \geq q$. Then ζ_q is q-primary, and we have $k(\sqrt[q]{\zeta_q}) = k(\sqrt[q]{\alpha_0})$. Therefore

$$\zeta_q = \alpha_0^x \alpha^q, \quad \alpha \in k, \; p \nmid x.$$

As before we find

$$x \equiv \frac{g-1}{q} \bmod q.$$

Again we may assume without loss of generality that $x = \frac{g-1}{q}$, and since α_0 is determined only up to q-th powers, we may change α_0 in such a way that

$$\zeta_q = \alpha_0^{(g-1)/q} \alpha_1^{q'},$$

where α_1 cannot be written in the form $\alpha_0^a \beta^p$ for $a \in \mathbb{Z}$ and $\beta \in k$.

Because of (8.5) and (v), a uniformizer π of k satisfies

$$(\alpha_0, \pi) \sqrt[q]{\alpha_0} = (\pi, k(\sqrt[q]{\alpha_0})/k) \sqrt[q]{\alpha_0} = \zeta_q \sqrt[q]{\alpha_0},$$

i.e., $(\alpha_0, \pi) = \zeta_q$.

For a unit α' of k we get accordingly $(\alpha_0, \alpha') = 1$.

There exists a unit α' such that $(\alpha_1, \alpha') - \zeta_q$: Suppose we have $(\alpha_1, \alpha') = \zeta_q^{pb}$ for all α'. Then we consider $\gamma = (\alpha_1 \alpha_0^{-c})^{q/p}$, where c satisfies $(\alpha_1, \pi) = \zeta_q^c$. Then γ is not a q-th power by construction, and we have $(\beta, \gamma) = 1$ for all $\beta \in k^\times$, contradicting Section 8.11, (iv).

Thus we can change π by a unit in such a way that $(\pi, \alpha_1) = 1$, and then $\alpha_1 k^{\times q}$ is an element of the $\mathbb{Z}/q$-module M formed by all $\alpha k^{\times q}$ with $(\pi, \alpha) = (\alpha_0, \alpha) = 1$. Since $(\alpha_0, \alpha) = 1$, α is a unit. The norm residue symbol defines by Section 8.11, (iv) a nondegenerate bilinear form on M. This implies that we can choose $\alpha_2, \ldots, \alpha_n$ in such a way that the conditions (iii) and (iv) are satisfied,

The elements $\pi, \alpha_0, \alpha_1, \ldots, \alpha_n$ form a basis of $k^\times / k^{\times q}$ and satisfy the conditions (i) – (vi). □

Lemma 10.7. *Let $\{\alpha_1, \ldots, \alpha_m\}$ be an orthonormal basis of $k^\times / k^{\times q}$, i.e., let*

$$(\alpha_{2\nu-1}, \alpha_{2\nu}) = \zeta_q, \quad \nu = 1, \ldots, m/2,$$

and $(\alpha_\nu, \alpha_\mu) = 1$ for all other pairs of basis elements α_ν, α_μ, $\nu < \mu$.

Assume moreover that

$$(\alpha_\nu, \zeta_q) = \zeta_q^{a_\nu}, \quad \nu = 1, \ldots, m,$$

and that σ_ν is an automorphism of $\widehat{k}/k$ with

$$\sigma_\nu \sqrt[q]{\alpha_\nu} = \zeta_q \sqrt[q]{\alpha_\nu} \tag{10.7}$$

$$\sigma_\nu \sqrt[q]{\alpha_\mu} = \sqrt[q]{\alpha_\mu}, \quad \mu \neq \nu, \mu = 1, \ldots, m. \tag{10.8}$$

Then $\{\sigma_1, \ldots, \sigma_m\}$ is a minimal system of generators of G_k Let

$$1 \longrightarrow R \longrightarrow F \longrightarrow G_k \longrightarrow 1$$

be a minimal presentation of G_k, and let $s_1, \ldots, s_m$ be preimages of $\sigma_1, \ldots, \sigma_m$ in F. The unique generating relation r of G_k can be chosen in the form

$$r = \Big(\prod_{\nu=1}^{m} s_\nu^{a_\nu q}\Big)[s_1, s_2] \cdots [s_{m-1}, s_m] r'$$

with $r' \in F^{(3,q)}$.

Proof. The elements $\sigma_1 G_k^{(2,q)}, \ldots, \sigma_m G_k^{(2,q)}$ form a minimal system of generators of $G_k/G_k^{(2,q)}$. By Theorem 4.10, we conclude that $\{\sigma_1, \ldots, \sigma_m\}$ is a minimal system of generators of G_k.

Let $\chi_1, \ldots, \chi_m$ be the dual basis of the basis $\{\sigma_1, \ldots, \sigma_m\}$ of $\mathrm{H}^1(G_k, \mathbb{Z}/q)$, i.e.,

$$\chi_\nu(\sigma_\mu) = \delta_{\nu\mu}, \quad \nu, \mu = 1, \ldots, m.$$

By Theorems 8.12 and 8.13, we get

$$\chi_{2\nu-1} \cup \chi_{2\nu} = -\xi_0, \quad \nu = 1, \ldots, m/2,$$

and for all other pairs of basis elements χ_ν, χ_μ with $\nu < \mu$ the cup product equals 0. Here $\xi_0 = \psi^{-1}(\zeta_q)$ generates the cyclic group $\mathrm{H}^2(G_k, \mathbb{Z}/q)$ of order q.

By Theorem 10.3, G_k has relation rank 1. The transgression induces an isomorphism $\mathrm{H}^1(R, \mathbb{Z}/q)^{G_k} \simeq \mathrm{H}^2(G_k, \mathbb{Z}/q)$. Let r be a generating relation with

$$\mathrm{tra}^{-1}\xi_0(r) = -1.$$

Then we get

$$r = \left(\prod_{\nu=1}^{m} s_\nu^{a_\nu q}\right)[s_1, s_2] \cdots [s_{m-1}, s_m] r', \quad r' \in F^{(3,q)}$$

from Theorems 7.23 and 7.24. □

Lemma 10.8. *Let the notation be as in Lemma 10.7, and view $\sigma_1, \ldots, \sigma_m$ as automorphisms of $k^{(2,q)} = k(\sqrt[q]{\alpha_1}, \ldots, \sqrt[q]{\alpha_m})$. Then we have*

$$(\alpha_{2\nu-1}, k^{(2,q)}/k) = \sigma_{2\nu}^{-1} \sigma_{2\nu-1}^{\binom{q}{2} a_{2\nu-1}}, \tag{10.9}$$

$$(\alpha_{2\nu}, k^{(2,q)}/k) = \sigma_{2\nu-1} \sigma_{2\nu}^{\binom{q}{2} a_{2\nu}}, \quad \nu = 1, \ldots, m/2. \tag{10.10}$$

Proof. We first remark that the norm of $\sqrt[q]{\alpha}$ in $k(\sqrt[q]{\alpha'})/k$ equals $\zeta_q^{\binom{q}{2}} \alpha$. Therefore we get

$$\left(\alpha, \zeta_q^{\binom{q}{2}} \alpha\right) = 1,$$

and hence

$$(\alpha_\nu, \alpha_\nu) = \left(\alpha_\nu, \zeta_q^{\binom{q}{2}}\right) = \zeta_q^{\binom{q}{2} a_\nu}.$$

Moreover, for $\mu = 1, \ldots, m$ and $\nu = 1, \ldots, m/2$ we have

$$\begin{aligned}(\alpha_{2\nu-1}, k^{(2,q)}/k) \sqrt[q]{\alpha_\mu} &= (\alpha_{2\nu-1}, k(\sqrt[q]{\alpha_\mu})/k) \sqrt[q]{\alpha_\mu} = (\alpha_\mu, \alpha_{2\nu-1}) \sqrt[q]{\alpha_\mu} \\ &= \sigma_{2\nu}^{-1} \sigma_{2\nu-1}^{\binom{q}{2} a_{2\nu-1}} \sqrt[q]{\alpha_\mu}, \\ (\alpha_{2\nu}, k^{(2,q)}/k) \sqrt[q]{\alpha_\mu} &= (\alpha_{2\nu}, k(\sqrt[q]{\alpha_\mu})/k) \sqrt[q]{\alpha_\mu} = (\alpha_\mu, \alpha_{2\nu}) \sqrt[q]{\alpha_\mu} \\ &= \sigma_{2\nu-1}^{-1} \sigma_{2\nu}^{\binom{q}{2} a_{2\nu}} \sqrt[q]{\alpha_\mu}.\end{aligned}$$

□

Now it is easy to prove the following theorem.

Theorem 10.9. *Let k be a finite extension of $\mathbb{Q}_p$ of even degree n, and let $q \geq p$ $(\widehat{q})$ denote the maximal p-power such that k (the maximal unramified p-extension T_k) contains the q-th ($\widehat{q}$-th) roots of unity. Let $\zeta_{\widehat{q}}$ be a primitive $\widehat{q}$-th root of unity,*

$$\zeta_q = \zeta_{\widehat{q}}^{q'}, \quad q' = \widehat{q}q^{-1},$$

and let g be an integer such that

$$\sigma' \zeta_{\widehat{q}} = \zeta_{\widehat{q}}^{g},$$

where σ' is the Frobenius automorphism of T_k/k.

Moreover, let $\{\pi, \alpha_0, \alpha_1, \ldots, \alpha_n\}$ be a basis of $k^\times/k^{\times q}$ satisfying the conditions (i) – (vi) of Lemma 10.6, σ an arbitrary lift of σ' to $\widehat{k}$ such that

$$\sigma G_k^{(2)} = (\pi, k^{(2)}/k),$$

and $\tau_0, \ldots, \tau_n$ arbitrary elements of G such that

$$\tau_\nu G_k^{(2)} = (\alpha_\nu, k^{(2)}/k), \quad \nu = 0, 1, \ldots, n,$$

where $k^{(2)}$ is the maximal abelian extension of k.

Then $\tau_0, \ldots, \tau_n$ are elements of the inertia group of $\widehat{k}/k$ that, together with σ, form a minimal system of generators of G_k. Let

$$1 \longrightarrow R \longrightarrow F \longrightarrow G_k \longrightarrow 1$$

be a minimal presentation of G_k, and let s_0, t_0, ..., t_n denote preimages of σ, τ_0, ..., τ_n in F. The unique generating relation r of G_k can be chosen in the form

$$r = t_0^{g-1} t_1^{\widehat{q}} [t_0, s][t_1, t_2] \cdots [t_{n-1}, t_n] r', \quad r' \in F^{(3,q)} \cap F^{(2)}.$$

Proof. Since $k^{(2)}$ contains the maximal unramified subextension T_k/k of $\widehat{k}/k$, by Section 8.6 τ_0, ..., τ_n are elements of $\mathrm{Gal}(\widehat{k}/T_k)$.

We now put $\alpha_0 = \alpha_{n+1}$, $\pi = \alpha_{n+2}$, $n + 2 = m$ and apply Lemma 10.7.

We first compute a_1, ..., a_m. By Lemma 10.6 we have $\zeta_q = \alpha_0^{(g-1)/q} \alpha_1^{q'}$, hence

$$\begin{aligned}
(\alpha_\nu, \zeta_q) &= 1 \quad \text{for } \nu = 3, 4, \ldots, m-1, \\
(\pi, \zeta_q) = \left(\pi, \alpha_0^{(g-1)/q}\right) &= \zeta_q^{-(g-1)/q}, \\
(\alpha_2, \zeta_q) = \left(\alpha_2, \alpha_1^{q'}\right) &= \zeta_q^{-q'}.
\end{aligned}$$

It remains to compute (α_1, ζ_q). To this end we show that $(\zeta_q, \zeta_q) = 1$.

For $p \neq 2$ we find

$$(\zeta_q, \zeta_q) = (\zeta_q, \zeta_q)^{-1} = 1.$$

For $p = 2$, $q \geq 4$, raising $(-\zeta_q, \zeta_q) = 1$ to the q-th power gives

$$(-1, \zeta_q) = 1$$

and thus $(\zeta_q, \zeta_q) = 1$. For $p = q = 2$, we get by Section 8.6, (i)

$$\begin{aligned}(-1,-1)\sqrt{-1} &= (-1, k(\sqrt{-1})/k)\sqrt{-1} \\ &= (N_{k/\mathbb{Q}_2}(-1), \mathbb{Q}_2(\sqrt{-1})/\mathbb{Q}_2)\sqrt{-1} \\ &= \sqrt{-1}\end{aligned}$$

and therefore $(-1,-1) = 1$, since by assumption $(k : \mathbb{Q}_2) = n$ is even.

For $q' = 1$ we get

$$(\alpha_1, \zeta_q) = (\zeta_q, \zeta_q) = 1,$$

and for $q' \geq p$

$$(\alpha_1, \zeta_q) = (\alpha_1, \alpha_1)^{q'} = 1.$$

This implies

$$\tilde{s}a_2 = -q', \quad a_m = -\frac{g-1}{q}, \quad a_1 = a_3 = \ldots = a_{m-1} = 0.$$

By Lemma 10.8 we can choose $\sigma_1, \ldots, \sigma_m$ in such a way that the relations

$$\begin{aligned}\tau_0 = \sigma_m^{-1}, \quad \sigma &= \sigma_{m-1}\sigma_m^{-\binom{q}{2}\frac{g-1}{q}}, \\ \tau_1 = \sigma_2^{-1}, \quad \tau_2 &= \sigma_1\sigma_2^{-\binom{q}{2}q'}, \\ \tau_{2\nu-1} = \sigma_{2\nu}^{-1}, \tau_{2\nu} &= \sigma_{2\nu-1}, \quad \nu = 2, \ldots, n/2\end{aligned}$$

hold. This implies that $\sigma, \tau_0, \ldots, \tau_n$ is a minimal system of generators of G_k and that the generating relation r_1 of G_k has the form

$$r_1 = t_0^{g-1}t_1^{\hat{q}}\left[t_2 t_1^{\binom{q}{2}q'}, t_1^{-1}\right][t_4, t_3^{-1}][t_6, t_5^{-1}]\cdots[t_n, t_{n-1}^{-1}]\left[s t_0^{\binom{q}{2}\frac{g-1}{q}}, t_0^{-1}\right]r_2$$

for $r_2 \in F^{(3,q)}$.

This gives after an easy calculation

$$r_1 = t_0^{g-1}t_1^{\hat{q}}[t_0, s][t_1, t_2]\cdots[t_{n-1}, t_n]r_3, \quad r_3 \in F^{(3,q)}. \tag{10.11}$$

Since $\zeta_q = \alpha_0^{(g-1)/q}\alpha_1^{q'}$ and

$$(\zeta_q, k^{(2)}/k)^q = 1,$$

r_1 must have the form

$$r_1 = (t_0^{g-1}t_1^{\hat{q}})^{\kappa} r_4, \quad r_4 \in [F, F] = F^{(2)}, \ \kappa \in \mathbb{Z}_p.$$

Comparing with (10.11) gives $\kappa \equiv 1 \pmod{q}$. Since r_1 is a generating relation of G_k, so is $r_1^{\kappa-1} = r$, and r has the desired properties. □

The remaining case $n \equiv 1 \pmod 2$ is analogous to the one we just considered; we will therefore be brief.

We have $q = p - 2$, and the extension $k(\sqrt{-1})/k$ is ramified, since otherwise $k/\mathbb{Q}_2$ would have even ramification index. Moreover,

$$(-1, -1) = -1.$$

In fact, we have

$$\begin{aligned}(-1,-1)\sqrt{-1} &= (-1, k(\sqrt{-1})/k)\sqrt{-1} \\ &= (N_{k/\mathbb{Q}_2}(-1), \mathbb{Q}_2(\sqrt{-1})/\mathbb{Q}_2) \\ &= (-1, \mathbb{Q}_2(\sqrt{-1})/\mathbb{Q}_2)\sqrt{-1} \;=\; -\sqrt{-1},\end{aligned}$$

since -1 is not a norm in $\mathbb{Q}_2(\sqrt{-1})/\mathbb{Q}_2$. This proves

Lemma 10.10. *There is a basis $\{\pi, \alpha_0, \alpha_1, \ldots, \alpha_n\}$ of $k^\times/k^{\times 2}$ satisfying the following conditions:*

(i) *π is a uniformizer of k;*

(ii) *α_0 is 2-primary;*

(iii) *$\alpha_0, \alpha_1, \ldots, \alpha_n$ are units in k.*

(iv) *$(\alpha_0, \pi) = -1$, $(\alpha_1, \pi) = -1$,*

$$(\alpha_{2\nu}, \alpha_{2\nu+1}) = -1, \quad \nu = 1, \ldots, \tfrac{n-1}{2};$$

for all other pairs of basis elements the norm residue symbols equal 1.

(v) *$-1 = \alpha_1$.*

The following Lemma corresponds to Lemma 10.7:

Lemma 10.11. *Let the notation be as in Lemma 10.10. Put $\alpha_{n+1} = \alpha_0$, $\alpha_{n+2} = \pi$, and let $\sigma_1, \ldots, \sigma_m$ be automorphisms of $k^{(2)}/k$ such that*

$$\sigma_\nu(\sqrt{\alpha_\mu}) = (-1)^{\delta_{\nu\mu}}\sqrt{\alpha_\mu}, \quad \nu, \mu = 1, \ldots, m.$$

Then

$$\begin{aligned}(\alpha_1, k^{(2)}/k) &= \sigma_1, \\ (\alpha_{2\nu}, k^{(2)}/k) &= \sigma_{2\nu+1}, \\ (\alpha_{2\nu+1}, k^{(2)}/k) &= \sigma_{2\nu}, \quad \nu = 1, \ldots, \frac{n-1}{2}.\end{aligned}$$

In analogy to Theorem 10.9 we now can prove

Theorem 10.12. *Let k be a finite extension of $\mathbb{Q}_2$ with odd degree n. Let $\{\pi, \alpha_0, \ldots, \alpha_n\}$ be a basis of $k^\times/k^{\times 2}$ satisfying the conditions (i)–(v) of Lemma 10.10; finally, let $\sigma, \tau_0, \ldots, \tau_n$ be arbitrary elements of G_k such that*

$$\sigma G_k^{(2)} = (\pi, k^{(2)}/k),$$
$$\tau_\nu G_k^{(2)} = (\alpha_\nu, k^{(2)}/k), \quad \nu = 0, 1, \ldots, n.$$

Then σ is a lift of the Frobenius automorphism of T_k/k, $\tau_0, \ldots, \tau_n$ are elements of the inertia group of $\widehat{k}/k$ and together with σ form a minimal system of generators of G_k. Let

$$1 \longrightarrow R \longrightarrow F \longrightarrow G \longrightarrow 1$$

be a minimal presentation of G, and let $s, t_0, \ldots, t_n$ be the preimages of σ, $\tau_0, \ldots, \tau_n$ in F. The single generating relation r of G can be chosen in the form

$$r = t_1^2[t_0, s][t_2, t_3] \cdots [t_{n-1}, t_n]r', \quad r' \in F^{(3,2)} \cap F^{(2)}.$$

11. Global Fields of Finite Type

We now consider as our base fields global fields k of finite type, i.e., finite extensions $k/\mathbb{Q}$ or $k/\mathbb{F}_q(x)$. The results of this chapter form the central part of this book. We are going to study the Galois group G_S of the maximal p-extension of k unramified outside some set of primes S of k, and we investigate in particular how to use the theorems on the structure of the Galois group of the maximal p-extension of a local field derived in the last chapter to get results on G_S.

11.1 The Maximal p-Extension

Let $\mathfrak{p}$ be a prime of k, and let $k_\mathfrak{p}$ denote the completion of k with respect to $\mathfrak{p}$. We put $G = \mathrm{Gal}(\widehat{k}/k)$ and $G_\mathfrak{p} = \mathrm{Gal}(\widehat{k_\mathfrak{p}}/k_\mathfrak{p})$. An embedding of $\widehat{k}$ in $\widehat{k_\mathfrak{p}}$ corresponds to the morphism $\varphi_\mathfrak{p} : G_\mathfrak{p} \longrightarrow G$ assigning to every automorphism of $G_\mathfrak{p}$ its restriction to $\widehat{k}$. Then $\varphi_\mathfrak{p}$ induces a homomorphism

$$\varphi_\mathfrak{p}^* : \mathrm{H}^\nu(G) \longrightarrow \mathrm{H}^\nu(G_\mathfrak{p}).$$

The homomorphism $\varphi_\mathfrak{p}^*$ does not depend on the choice of the embedding $\psi_\mathfrak{p} : \widehat{k} \hookrightarrow \widehat{k_\mathfrak{p}}$. If $\psi'_\mathfrak{p} : \widehat{k} \hookrightarrow \widehat{k_\mathfrak{p}}$ is another embedding, then by the lifting theorem for isomorphisms there is an automorphism $\gamma : \widehat{k_\mathfrak{p}} \longrightarrow \widehat{k_\mathfrak{p}}$ making the diagram

$$\begin{array}{ccc} \widehat{k} & \xrightarrow{\ \psi_\mathfrak{p}\ } & \widehat{k_\mathfrak{p}} \\ & \underset{\psi'_\mathfrak{p}}{\searrow} & \big\downarrow \gamma \\ & & \widehat{k_\mathfrak{p}} \end{array} \tag{11.1}$$

commutative. To this diagram (11.1) there corresponds a commutative diagram

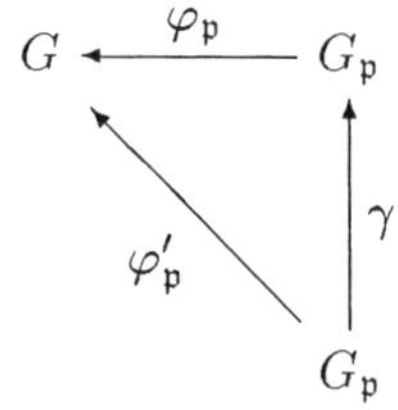

where γ now denotes the map $\tau \longmapsto \gamma^{-1}\tau\gamma$, $\tau \in G_{\mathfrak{p}}$. By Theorem 3.13, the induced map $\gamma^* : \mathrm{H}^{\nu}(G_{\mathfrak{p}}) \longrightarrow \mathrm{H}^{\nu}(G_{\mathfrak{p}})$ is the identity map: this implies the claim.

For finite primes $\mathfrak{p}$, the quotient group of $G_{\mathfrak{p}}$ modulo the inertia subgroup $\mathcal{T}_{\mathfrak{p}}$ of $\widehat{k}_{\mathfrak{p}}/k_{\mathfrak{p}}$ is isomorphic to $\mathbb{Z}_p$. For every $\mathfrak{p}$ we fix a place $\mathfrak{P}$ in $\widehat{k}$ above $\mathfrak{p}$. Then we have a canonical embedding $\widehat{k} \hookrightarrow \widehat{k}_{\mathfrak{P}}$ (see Section 8.1), and since $\widehat{k}_{\mathfrak{P}}/k_{\mathfrak{p}}$ is a p-extension, it can be embedded in $\widehat{k_{\mathfrak{p}}}$. In the following, we let $\varphi_{\mathfrak{p}} : G_{\mathfrak{p}} \longrightarrow G$ denote the morphism induced by the embedding $\widehat{k} \longrightarrow \widehat{k}_{\mathfrak{P}} \longrightarrow \widehat{k_{\mathfrak{p}}}$.

Let V be the set of all places of k. The family $\{\varphi_{\mathfrak{p}} \,|\, \mathfrak{p} \in V\}$ is admissible with respect to $\{\mathcal{T}_{\mathfrak{p}} \,|\, \mathfrak{p} \in V\}$ in the sense of section 6.2: let U be an open subgroup of G. The fixed field $K(U)$ of U is a finite extension of k, hence the number of primes ramified in $K(U)/k$ is finite. On the other hand, the image $\varphi_{\mathfrak{p}}(\mathcal{T}_{\mathfrak{p}})$ is the inertia subgroup $\mathcal{T}_{\mathfrak{P}}$ of $\mathfrak{P}$ in $\widehat{k}$. The fixed field $K(\mathcal{T}_{\mathfrak{P}})$ is the maximal subfield of $\widehat{k}$ in which $\mathfrak{P}$ is unramified. Thus $K(U) \subseteq K(\mathcal{T}_{\mathfrak{P}})$ for almost all $\mathfrak{p}$, and so $\varphi_{\mathfrak{p}}(\mathcal{T}_{\mathfrak{p}}) \subseteq U$.

Therefore the map

$$\varphi^* : \mathrm{H}^2(G) \longrightarrow \coprod_{\mathfrak{p}} \mathrm{H}^2(G_{\mathfrak{p}}),$$

where the sum is over all places $\mathfrak{p}$ of k, is well defined. For an infinite place $\mathfrak{p}$, we have $\mathrm{H}^2(G_{\mathfrak{p}}) \neq 0$ only if $p = 2$ and $\mathfrak{p}$ is real.

Theorem 11.1. *The map φ^* is injective.*

Proof. If the characteristic of k equals p, Theorem 11.1 holds because of Theorem 9.1. Assume therefore that $\chi(k) \neq p$, and consider the case $\delta(k) = 1$ first, i.e., k contains a primitive p-th root ζ_p of unity. The diagram of pairs

$$\begin{array}{ccc} \{G, \mathbb{F}_p\} & \longrightarrow & \{G_{\mathfrak{p}}, \mathbb{F}_p\} \\ \downarrow & & \downarrow \\ \{G, \widehat{k}^{\times}\} & \longrightarrow & \{G_{\mathfrak{p}}, \widehat{k}_{\mathfrak{p}}^{\times}\}, \end{array}$$

where the horizontal arrows are induced by the embedding $\widehat{k} \hookrightarrow \widehat{k}_{\mathfrak{p}}$ and the vertical arrows by an isomorphism $\mathbb{F}_p \longrightarrow \mu_p$, induces a commutative diagram

$$\begin{array}{ccc} \mathrm{H}^2(G) & \longrightarrow & \coprod_{\mathfrak{p}} \mathrm{H}^2(G_{\mathfrak{p}}) \\ \downarrow & & \downarrow \\ \mathrm{H}^2(G,\widehat{k}^\times)[p] & \longrightarrow & \coprod_{\mathfrak{p}} \mathrm{H}^2(G_{\mathfrak{p}},\widehat{k}_{\mathfrak{p}}^\times)[p] \end{array} \tag{11.2}$$

The lower horizontal arrow in (11.2) is well defined and injective: first, by Hasse's Local-Global Principle (Section 8.9), we have an injection

$$\mathrm{H}^2(G,\widehat{k}^\times) \hookrightarrow \coprod_{\mathfrak{p}} \mathrm{H}^2(\mathcal{Z}_{\mathfrak{P}}, k_{\mathfrak{P}}^\times),$$

where $\mathcal{Z}_{\mathfrak{P}}$ denotes the Galois group of $\widehat{k}_{\mathfrak{P}}/k_{\mathfrak{p}}$. Second, the homomorphism

$$\varphi_{\mathfrak{P}}^* : \mathrm{H}^2(\mathcal{Z}_{\mathfrak{P}},\widehat{k}_{\mathfrak{P}}^\times)[p] \longrightarrow \mathrm{H}^2(G_{\mathfrak{p}},\widehat{k}_{\mathfrak{p}}^\times)[p]$$

induced by the embedding $\widehat{k}_{\mathfrak{P}} \hookrightarrow \widehat{k}_{\mathfrak{p}}$ is an isomorphism: in fact, by Section 8.9 $\varphi_{\mathfrak{P}}^*$ is a monomorphism of cyclic groups of order dividing p. For finite $\mathfrak{p}$, the extension $\widehat{k}_{\mathfrak{P}}/k_{\mathfrak{p}}$ is always nontrivial, hence $\#\mathrm{H}^2(\mathcal{Z}_{\mathfrak{P}}, k_{\mathfrak{P}}^\times)[p] = p$. For infinite $\mathfrak{p}$, we have $\widehat{k}_{\mathfrak{P}} = \widehat{k}_{\mathfrak{p}} = \mathbb{C}$.

By Theorem 9.2, the vertical arrows in (11.2) are isomorphisms, and this implies the claim.

Now assume that $\delta(k) = 0$. We put $k' = k(\zeta_p)$ and $G' = \mathrm{Gal}(\widehat{k'}/k')$. The natural morphisms $G' \longrightarrow G$ and $G_{\mathfrak{p}'} \longrightarrow G_{\mathfrak{p}}$, where $\mathfrak{p}'$ is a prime divisor of k' above $\mathfrak{p}$, give rise to a commutative diagram

$$\begin{array}{ccc} \mathrm{H}^2(G) & \xrightarrow{\varphi^*} & \coprod_{\mathfrak{p}} \mathrm{H}^2(G_{\mathfrak{p}}) \\ \downarrow & & \downarrow \\ \mathrm{H}^2(G') & \xrightarrow{\varphi'^*} & \coprod_{\mathfrak{p}'} \mathrm{H}^2(G_{\mathfrak{p}'}) \end{array} \tag{11.3}$$

where $\mathfrak{p}'$ runs through all primes of k' and where $\theta_{\mathfrak{p}'} : \mathrm{H}^2(G_{\mathfrak{p}}) \longrightarrow \mathrm{H}^2(G_{\mathfrak{p}'})$ maps $\alpha_{\mathfrak{p}} \in \mathrm{H}^2(G_{\mathfrak{p}})$ to

$$\sum_{\mathfrak{p}'|\mathfrak{p}} \theta_{\mathfrak{p}'}(\alpha_{\mathfrak{p}}).$$

By Section 9.3, the vertical arrows in (11.3) are injective. Thus if φ'^* is injective, then so is φ^*. □

From Theorem 6.14 it can be seen that Theorem 11.1 means that the relation module of G is generated by the "local relations" for each $\mathfrak{p}$, of which there is one or none according as $\delta(k_{\mathfrak{p}}) = 1$ or $= 0$. If $\delta(k_{\mathfrak{p}}) = 1$, then according to the next theorem we can omit one of the local relations:

Theorem 11.2. *Let $\mathfrak{q}$ be an arbitrary prime of k and*

$$\varphi_{\mathfrak{q}}^* : \mathrm{H}^2(G) \longrightarrow \coprod_{\mathfrak{p}\neq\mathfrak{q}} \mathrm{H}^2(G_{\mathfrak{p}})$$

the map one gets from φ^* by omitting the summand $\mathrm{H}^2(G_{\mathfrak{q}})$ in the target group. If $\delta(k) = 1$, then $\varphi_{\mathfrak{q}}^*$ is injective.

Proof. We only have to remark that, by Hasse's Local-Global Principle, the map

$$\mathrm{H}^2(G, \widehat{k}^\times)[p] \longrightarrow \coprod_{\mathfrak{p}\neq\mathfrak{q}} \mathrm{H}^2(G_{\mathfrak{p}}, \widehat{k}_{\mathfrak{p}}^\times)[p]$$

is still injective. The proof of Theorem 11.1 therefore carries over. □

11.2 The Maximal p-Extension with Restricted Ramification

Let S be an arbitrary set of primes of k. We denote by k_S the maximal p-extension of k unramified outside S, i.e., ramified only at primes in S. Then k_S is the compositum of all finite p-extensions unramified outside S.

The following type of primes $\mathfrak{p}$ cannot ramify in p- extensions:

1. primes $\mathfrak{p}$ with $N(\mathfrak{p}) \not\equiv 0, 1 \pmod p$,
2. complex places,
3. real places for $p \neq 2$.

In the following we assume that S does not contain such primes.

Next we want to carry over Theorems 11.1 and 11.2 to $G_S = \mathrm{Gal}\,(k_S/k)$. The map corresponding to φ^* is

$$\varphi_S^* : \mathrm{H}^2(G_S) \longrightarrow \coprod_{\mathfrak{p}\in S} \mathrm{H}^2(G_{\mathfrak{p}})$$

induced by the morphism $\varphi_{\mathfrak{p}}^S : G_{\mathfrak{p}} \longrightarrow G \longrightarrow G_S$. Let Ш_S denote the kernel of φ_S^*.

Our task is to give a good approximation of the group Ш_S. To this end we define

$$V_S = \{\alpha \in k^\times \mid (\alpha) = \mathfrak{a}^p,\ \alpha \in k_{\mathfrak{p}}^p \text{ for } \mathfrak{p} \in S\},$$

where (α) is the principal ideal generated by α. We clearly have $k^{\times p} \subseteq V_S$. Thus we can put

$$\text{Б}_S = (V_S/k^{\times p})^*;$$

then Б_S is a finite group of exponent dividing p. For a proof, assume first that $S = \varnothing$ is the empty set. The homomorphism assigning to each $\alpha \in V_\varnothing$ the ideal $\mathfrak{a}$ with $(\alpha) = \mathfrak{a}^p$ induces an epimorphism $V_\varnothing/k^{\times p} \longrightarrow \mathrm{Cl}\,(k)[p]$ whose kernel is isomorphic to E/p, where E denotes the unit group of k. This shows that

$$\dim \text{Б}_\varnothing = \dim \mathrm{Cl}\,(k)[p] + \dim E/p$$

is finite. Moreover,

$$V_{S_1} \subseteq V_{S_2}$$

for $S_1 \supseteq S_2$. This implies the claim.

We now want to prove the following theorem which represents the main result of this book.

Theorem 11.3. *There is a natural injection* $\text{Ш}_S \hookrightarrow \text{Б}_S$.

Remark. By Theorem 6.11, the content of Theorem 11.3 is the following: for $\mathfrak{p} \in S$. Let

$$\begin{array}{ccccc} R_S & \longrightarrow & F_S & \longrightarrow & G_S \\ \uparrow & & \chi_\mathfrak{p}\uparrow & & \varphi_\mathfrak{p}\uparrow \\ R_\mathfrak{p} & \longrightarrow & F_\mathfrak{p} & \longrightarrow & G_\mathfrak{p} \end{array}$$

be a family of commutative diagrams that are admissible with respect to $\{\mathcal{T}_\mathfrak{P} \mid \mathfrak{p} \in S\}$ in the sense of theorem 6.7. Assume that the presentations of G_S and $G_\mathfrak{p}$ are minimal. According to Theorem 6.11, R_S is generated by the images under $\chi_\mathfrak{p}$ of the relations of $R_\mathfrak{p}$ for $\mathfrak{p} \in S$ (by Chapter 10, $R_\mathfrak{p}$ is generated by at most one relation) and by at most $\dim \text{Б}_S$ elements according to Theorem 11.3.

Proof of Theorem 11.3. For a prime $\mathfrak{p}$ of k let $\mathfrak{P}$ denote the prime divisor of $\mathfrak{p}$ in $\widehat{k}$ fixed in Section 11.1, and let $\mathcal{T}_S$ denote the normal subgroup of $G = \mathrm{Gal}\,(\widehat{k}/k)$ generated by the inertia subgroups $\mathcal{T}_\mathfrak{P}$ for $\mathfrak{p} \notin S$. Since the inertia groups of the other prime divisors of $\mathfrak{p}$ in $\widehat{k}$ are conjugates of $\mathcal{T}_\mathfrak{P}$, Theorem 8.3 implies that the fixed field of $\mathcal{T}_S$ equals k_S. We therefore have an exact sequence

$$1 \longrightarrow \mathcal{T}_S \longrightarrow G \longrightarrow G_S \longrightarrow 1. \tag{11.4}$$

By applying Theorem 3.14 to (11.4) and the module $\mathbb{F}_p$, we get the exact sequence

$$\mathrm{H}^1(G) \xrightarrow{\text{res}} \mathrm{H}^1(\mathcal{T}_S)^{G_S} \xrightarrow{\text{tra}} \mathrm{H}^2(G_S) \xrightarrow{\text{inf}} \mathrm{H}^2(G). \tag{11.5}$$

On the other hand, the diagram

$$\begin{array}{ccc} \mathrm{H}^2(G_S) & \longrightarrow & \mathrm{H}^2(G) \\ \varphi_S^\bullet\downarrow & & \downarrow\varphi^\bullet \\ \coprod_{\mathfrak{p}\in S} \mathrm{H}^2(G_\mathfrak{p}) & \xrightarrow{\iota} & \coprod_\mathfrak{p} \mathrm{H}^2(G_\mathfrak{p}), \end{array} \tag{11.6}$$

in which ι denotes the natural injection, commutes. In fact, for $\mathfrak{p} \notin S$ we have the commutative diagram

$$\begin{array}{ccc} G_{\mathfrak{p}} & \longrightarrow & G \\ \downarrow & & \downarrow \\ G_{\mathfrak{p}}/\mathcal{T}_{\mathfrak{p}} & \longrightarrow & G_S, \end{array}$$

where $\mathcal{T}_{\mathfrak{p}}$ denotes the inertia group of $\widehat{k}_{\mathfrak{p}}/k_{\mathfrak{p}}$. Now $G_{\mathfrak{p}}/\mathcal{T}_{\mathfrak{p}} \simeq \mathbb{Z}_p$ or $\simeq 1$ according as $\mathfrak{p}$ is finite or infinite. In any case, $\mathrm{H}^2(G_{\mathfrak{p}}/\mathcal{T}_{\mathfrak{p}}) = 0$, which gives the commutativity of (11.6).

By Theorem 11.1, φ^* is injective. Thus it follows from (11.6) that

$$\ker \inf = \ker \varphi_S^* = \text{Ш}_S.$$

By (11.5) we therefore have exact sequences

$$\mathrm{H}^1(G) \xrightarrow{\text{res}} \mathrm{H}^1(\mathcal{T}_S)^{G_S} \longrightarrow \text{Ш}_S \longrightarrow 0$$

and

$$0 \longrightarrow (\text{Ш}_S)^* \longrightarrow (\mathrm{H}^1(\mathcal{T}_S)^{G_S})^* \longrightarrow \mathrm{H}^1(G)^*.$$

Now

$$\mathrm{H}^1(G) = (G/\operatorname{Fr}(G))^*$$

and

$$\mathrm{H}^1(\mathcal{T}_S)^{G_S} = (\mathcal{T}_S/\mathcal{T}_S^p[\mathcal{T}_S, G])^*.$$

Therefore

$$0 \longrightarrow (\text{Ш}_S)^* \longrightarrow \mathcal{T}_S/\mathcal{T}_S^p[\mathcal{T}_S, G] \xrightarrow{\psi} G/\operatorname{Fr}(G)$$

is exact, where ψ is induced by the embedding $\mathcal{T}_S \hookrightarrow G$.

We now define a morphism

$$\chi : \prod_{\mathfrak{p} \notin S} \mathcal{T}_{\mathfrak{p}}/\mathcal{T}_{\mathfrak{p}}^p[\mathcal{T}_{\mathfrak{p}}, G_{\mathfrak{p}}] \longrightarrow \mathcal{T}_S/\mathcal{T}_S^p[\mathcal{T}_S, G].$$

To this end we consider the set $\mathfrak{M} = \{\mathcal{T}_{\mathfrak{P}} \mid \mathfrak{p} \notin S\}$. Every open subgroup U of G contains almost all groups $\mathcal{T}_{\mathfrak{P}}$ in $\mathfrak{M}$. In fact, the fixed field of U is a finite extension of k, in which therefore only finitely many primes ramify. For the unramified primes $\mathfrak{p}$ we have $\mathcal{T}_{\mathfrak{P}} \subseteq U$.

For $\tau_{\mathfrak{p}} \in \mathcal{T}_{\mathfrak{p}}$, we have $\varphi_{\mathfrak{p}}\tau_{\mathfrak{p}} \in \mathcal{T}_{\mathfrak{P}}$. Therefore

$$\prod_{\mathfrak{p} \notin S} \overline{\varphi_{\mathfrak{p}}\tau_{\mathfrak{p}}}$$

is a well defined element of $\mathcal{T}_S/\mathcal{T}_S^p[\mathcal{T}_S, G]$. Since $\varphi_{\mathfrak{p}}\tau_{\mathfrak{p}} \in \mathcal{T}_S^p[\mathcal{T}_S, G]$ for $\tau_{\mathfrak{p}} \in \mathcal{T}_{\mathfrak{p}}^p[\mathcal{T}_{\mathfrak{p}}, G_{\mathfrak{p}}]$, we can set

$$\chi\Bigl(\prod_{\mathfrak{p}\notin S}\overline{\tau}_{\mathfrak{p}}\Bigr) = \prod_{\mathfrak{p}\notin S}\overline{\varphi_{\mathfrak{p}}\tau_{\mathfrak{p}}}$$

and get the desired morphism χ.

Clearly the classes from $\mathcal{T}_S/\mathcal{T}_S^p[\mathcal{T}_S, G]$ having representatives in $\mathcal{T}_{\mathfrak{P}}$ are in im χ. Since these classes generate $\mathcal{T}_S/\mathcal{T}_S^p[\mathcal{T}_S, G]$, χ is surjective.

Since $G_{\mathfrak{p}}/\mathcal{T}_{\mathfrak{p}}$ is cyclic, we have $[\mathcal{T}_{\mathfrak{p}}, G_{\mathfrak{p}}] = [G_{\mathfrak{p}}, G_{\mathfrak{p}}]$. By local class field theory (Section 8.7), we have $\mathcal{T}_{\mathfrak{p}}/\mathcal{T}_{\mathfrak{p}}^p[\mathcal{T}_{\mathfrak{p}}, G_{\mathfrak{p}}] \simeq E_{\mathfrak{p}}/E_{\mathfrak{p}}^p$, where $E_{\mathfrak{p}}$ denotes the unit group of $k_{\mathfrak{p}}$. On the other hand, by Section 8.7 we also have $G/G^p[G,G] \simeq C_k/C_k^p = J/J^p k^\times$, where we have put $J = J(k)$. We therefore get a commutative diagram

$$\begin{array}{ccccc}
\prod_{\mathfrak{p}\notin S}\mathcal{T}_{\mathfrak{p}}/\mathcal{T}_{\mathfrak{p}}^p[\mathcal{T}_{\mathfrak{p}}, G_{\mathfrak{p}}] & \xrightarrow{\ \chi\ } & \mathcal{T}_S/\mathcal{T}_S^p[\mathcal{T}_S, G] & \xrightarrow{\ \psi\ } & G/G^p[G,G] \\
\downarrow & & & & \downarrow \\
\prod_{\mathfrak{p}\notin S} E_{\mathfrak{p}}/E_{\mathfrak{p}}^p & & \xrightarrow[\ \eta\ ]{} & & J/J^p k^\times
\end{array} \tag{11.7}$$

where η is induced by the embedding

$$\prod_{\mathfrak{p}\notin S} E_{\mathfrak{p}} \hookrightarrow J.$$

From (11.7) we get an isomorphism

$$\ker\eta \longrightarrow \ker\psi\chi,$$

and since χ is surjective, an epimorphism

$$\ker\psi\chi \longrightarrow \ker\psi.$$

We now compute $\ker\eta$: Let U_S be the group of all ideles of k whose component is 1 for $\mathfrak{p}\in S$ and a unit for $\mathfrak{p}\notin S$. (For an infinite prime $\mathfrak{p}$, all elements of $k_{\mathfrak{p}}^\times$ are regarded as units). Then

$$\prod_{\mathfrak{p}\notin S} E_{\mathfrak{p}}/E_{\mathfrak{p}}^p \simeq U_S J^p/J^p,$$

and

$$\ker\eta \simeq U_S J^p \cap J^p k^\times/J^p.$$

Since

$$U_S J^p \cap J^p k^\times = (U_S J^p \cap k^\times) J^p,$$

we find

$$\ker \eta \simeq (U_S J^p \cap k^\times)/(J^p \cap k^\times) = V_S/k^{\times p}.$$

Collecting everything we have found an epimorphism

$$V_S/k^{\times p} \longrightarrow (\text{Ш}_S)^*$$

and hence a monomorphism

$$\text{Ш}_S \longrightarrow \text{Б}_S.$$

□

We now return to the case $\delta(k) = 1$. Let $\mathfrak{q} \in S$, $S' = S \setminus \{\mathfrak{q}\}$, and let $\varphi^*_{S'}$ be the map

$$\mathrm{H}^2(G_S) \longrightarrow \coprod_{\mathfrak{p} \in S'} \mathrm{H}^2(G_\mathfrak{p}).$$

Theorem 11.4. *For $\delta(k) = 1$, we have*

$$\text{Ш}_S = \ker \varphi^*_{S'}.$$

Proof. We have a commutative diagram

$$\begin{array}{ccc} \mathrm{H}^2(G_S) & \xrightarrow{\text{inf}} & \mathrm{H}^2(G) \\ \varphi^*_{S'} \downarrow & & \varphi^*_\mathfrak{q} \downarrow \\ \coprod_{\mathfrak{p} \in S'} \mathrm{H}^2(G_\mathfrak{p}) & \longrightarrow & \coprod_{\mathfrak{p} \neq \mathfrak{q}} \mathrm{H}^2(G_\mathfrak{p}) \end{array}$$

Since $\varphi^*_\mathfrak{q}$ is injective, we have

$$\text{Ш}_S = \ker \inf = \ker \varphi^*_{S'}.$$

□

As a conclusion of Theorems 11.3 and 11.4, we get

Theorem 11.5. *Let S be a finite set of primes in k. Then we have the following estimate for the relation rank $r(G_S)$ of G_S:*

$$r(G_S) \leq \sum_{\mathfrak{p} \in S} \delta(k_\mathfrak{p}) - \delta(k) + \dim \text{Б}_S + \theta, \tag{11.8}$$

where $\theta = \theta(k, S) = 1$ for $\delta(k) = 1$ and $S = \varnothing$, and $\theta = 0$ otherwise.

Proof. By the definition of Ш_S and Theorem 11.4 we have

$$r(G_S) = \dim \mathrm{H}^2(G_S) \leq \dim \text{Ш}_S + \sum_{\mathfrak{p} \in S} \delta(k_\mathfrak{p}) - \delta(k) + \theta.$$

Moreover

$$\dim \text{Ш}_S \leq \dim \text{Б}_S$$

by Theorem 11.3, and this implies the claim. □

Example 11.6. If $\chi(k) = p$ and $\mathrm{Cl}\,(k)[p] = 1$, then G_S is a free pro-p group for all sets S of primes in k.

11.3 Generator Rank

The main aim in this section is the computation of the generator rank $d(G_S) = \dim \mathrm{H}^1(G_S)$ in the case where S is finite. We first construct a sequence which is exact for arbitrary S.

To this end we consider the following groups and morphisms: by definition, we have

$$\mathrm{H}^1(G_S) = (G_S/\operatorname{Fr}(G_S))^* \quad \text{with } \operatorname{Fr}(G_S) = G_S^p[G_S, G_S].$$

$G_S/\operatorname{Fr}(G_S)$ is the Galois group of the maximal abelian p-elementary extension unramified outside S. Class field theory (Sections 8.6, 8.7) yields an isomorphism

$$G_S/\operatorname{Fr}(G_S) \longrightarrow J/U_S J^p k^\times. \tag{11.9}$$

Let $U = U_\varnothing$. There is a natural isomorphism

$$U/U_S U^p \longrightarrow \prod_{\mathfrak{p}\in S} E_\mathfrak{p}/E_\mathfrak{p}^p. \tag{11.10}$$

From

$$V_S = U_S J^p \cap k^\times, \quad V_\varnothing/k^{\times p} \simeq U J^p \cap J^p k^\times / J^p$$

we get a natural sequence

$$\begin{array}{ccccccc} 0 & \longrightarrow & V_S/k^{\times p} & \xrightarrow{\varphi_1} & V_\varnothing/k^{\times p} & \xrightarrow{\varphi_2} & U/U_S U^p \\ & & & & & & \downarrow{\scriptstyle \varphi_3} \\ & & 0 & \longleftarrow & J/U J^p k^\times & \xleftarrow{\varphi_4} & J/U_S J^p k^\times \end{array} \tag{11.11}$$

that is easily checked to be exact. By observing (11.9), (11.10) and

$$(E_\mathfrak{p}/E_\mathfrak{p}^p)^* = (\mathcal{T}_\mathfrak{p}/\mathcal{T}_\mathfrak{p}[\mathcal{T}_\mathfrak{p}, G_\mathfrak{p}])^* = \mathrm{H}^1(\mathcal{T}_\mathfrak{p})^{G_\mathfrak{p}},$$

and taking duals we find

Theorem 11.7. *The sequence*

$$\begin{array}{ccccccc} 0 & \longrightarrow & \mathrm{H}^2(G_\varnothing) & \xrightarrow{\text{inf}} & \mathrm{H}^1(G_S) & \xrightarrow{\varphi_3^*} & \coprod_{\mathfrak{p}\in S} \mathrm{H}^1(\mathcal{T}_\mathfrak{p})^{G_\mathfrak{p}} \\ & & & & & & \downarrow{\scriptstyle \varphi_2^*} \\ & & 0 & \longleftarrow & \mathbf{Б}_S & \xleftarrow{\varphi_1^*} & \mathbf{Б}_\varnothing \end{array}$$

is exact. The map φ_3^* is induced by the morphisms

$$\mathcal{T}_\mathfrak{p} \longrightarrow G_\mathfrak{p} \xrightarrow{\varphi_\mathfrak{p}} G \longrightarrow G_S.$$

As we have already seen in Section 11.2,

$$\dim \mathbf{Б}_{\varnothing} = \dim \mathrm{Cl}\,(k)[p] + \dim E/p$$

is finite. On the other hand, $J/UJ^pk^\times$ is isomorphic to $\mathrm{Cl}\,(k)/p$. For algebraic number fields k, $\mathrm{Cl}\,(k)$ is finite and hence

$$\dim \mathrm{Cl}\,(k)[p] = \dim \mathrm{Cl}\,(k)/p.$$

For function fields k, however, we have an exact sequence

$$0 \longrightarrow \mathrm{Cl}_0(k) \longrightarrow \mathrm{Cl}\,(k) \xrightarrow{\deg} \mathbb{Z},$$

where $\mathrm{Cl}_0(k)$ denotes the group of divisor classes of degree 0, and where deg denotes the degree of the divisor class. Since $\mathrm{Cl}_0(k)$ is finite, we get

$$\mathrm{Cl}\,(k)[p] = \mathrm{Cl}_0(k)[p] \simeq \mathrm{Cl}_0(k)/p$$

and

$$\dim \mathrm{Cl}\,(k)/p = \dim \mathrm{Cl}_0(k)/p + 1 = \dim \mathrm{Cl}\,(k)[p] + 1.$$

For a function field, the unit group E consists only of roots of unity; therefore

$$\dim E/p = \delta(k).$$

For a number field k, we have

$$\dim E/p = \delta(k) + r_1 + r_2 - 1,$$

where r_1 is the number of real and r_2 half the number of complex conjugates of k.

For a local field $k_\mathfrak{p}$ of finite type, we have

$$\dim \mathrm{H}^1(\mathcal{T}_\mathfrak{p})^{G_\mathfrak{p}} = \dim E_\mathfrak{p}/E_\mathfrak{p}^p = \begin{cases} \delta(k_\mathfrak{p}) & \text{for } \chi(\mathfrak{p}) \neq p, \\ (k_\mathfrak{p} : \mathbb{Q}_p) + \delta(k_\mathfrak{p}) & \text{for } \chi(\mathfrak{p}) = p,\ \chi(k) = 0, \\ \infty & \text{for } \chi(k) = p. \end{cases}$$

by Section 8.5.

Collecting everything we now can determine the generator rank of G_S with the help of Theorem 11.7:

Theorem 11.8. *The rank $d(G_S)$ of generators of G_S is finite if and only if we are in one of the following cases:*

1. *$\chi(k) \neq p$ and the number of primes $\mathfrak{p}$ in S with $\delta(k_\mathfrak{p}) = 1$ is finite.*
2. *$\chi(k) = p$ and $S = \varnothing$.*

In both cases we have

$$d(G_S) = \sum_{\substack{\mathfrak{p}\in S \\ \chi(\mathfrak{p})=p}} (k_{\mathfrak{p}} : \mathbb{Q}_p) - \delta(k) - r + 1 + \sum_{\mathfrak{p}\in S} \delta(k_{\mathfrak{p}}) + \dim \textbf{Б}_S,$$

where r denotes the number of infinite primes of k.

Combining Theorem 11.5 with Theorem 11.8 we get the following inequality for the partial Euler-Poincaré characteristic defined in Section 5.2:

$$\chi_2(G_S) = 1 - d(G_S) + r(G_S) \leq - \sum_{\substack{\mathfrak{p}\in S \\ \chi(\mathfrak{p})=p}} (k_{\mathfrak{p}} : \mathbb{Q}_p) \; + r + \theta(k, S). \qquad (11.12)$$

The term

$$H(k,S) = - \sum_{\substack{\mathfrak{p}\in S \\ \chi(\mathfrak{p})=p}} (k_{\mathfrak{p}} : \mathbb{Q}_p) \; + r + \theta(k, S).$$

on the right hand side of (11.12) behaves multiplicatively in the following sense if $\theta(k,S) = 0$ and if S does not contain any infinite primes: let $\overline{k}/k$ be a finite subextension of k_S/k, and let $\overline{S}$ denote the set of primes in $\overline{k}$ above the primes in S. Then $\overline{k}_S = k_S$, and we have

$$H(\overline{k}, \overline{S}) = (\overline{k} : k) H(k, S).$$

With the help of Theorem 5.5 we now get

Theorem 11.9. *Let $\theta(k,S) = 0$, and assume that S contains no infinite primes. Assume moreover that $d(G_S)$ is finite, and that in (11.12) we have equality for all finite subextensions $\overline{k}/k$ of k_S/k. Then G_S is a pro-p group with cohomological dimension ≤ 2.*

Proof. Let H be a subgroup of finite index in G_S. The fixed field of H is a finite extension $\overline{k}$ of k contained in k_S. By the assumption in Theorem 11.9 we therefore get

$$\chi_2(H) = \chi_2(G_S) = \mathrm{H}(\overline{k}, \overline{S}) = (\overline{k} : k) H(k, S) = (G : H)\chi_2(G_S).$$

The claim now follows from Theorem 5.5. □

We shall return to the question of when there is equality in (11.12) in Chapter 13.

11.4 Explicit Computation of Generators and Relations

In this section we want to write down explicit expressions for the generating relations of G_S, where we restrict ourselves to a minimal system of generators of G_S that we have chosen in a "best possible" way. We only consider the case $\text{Б}_S = 0$, because this implies $\text{Ш}_S = 0$ by Theorem 11.3, which in turn means by Theorem 6.14 that the module of relations of G_S is generated by the local relations. If $\chi(k) = p$, then G_S is free in this case, so we may assume that $\chi(k) \neq p$.

We begin by choosing a minimal system of generators of G_S and introduce the following notation: Let $h = \dim \mathrm{Cl}\,(k)/p$ and let $i_1, \ldots, i_h$ be a system of ideles whose image under the map

$$J \longrightarrow J/UJ^p k^\times = \mathrm{Cl}\,(k)/p$$

form a basis of $\mathrm{Cl}\,(k)/p$. For function fields we assume that i_1 as minimal possible positive degree, and that the degrees of $i_2, \ldots, i_h$ are 0.

Let ζ be a generator of the group of roots of unity in k, and let $\varepsilon_1, \ldots, \varepsilon_{r-1}$ be a basis for the unit group of k. If $\delta(k) = 1$, then we put $\zeta = \varepsilon_r$.

For a prime $\mathfrak{p}$ in k let $\pi_\mathfrak{p}$ denote a uniformizer and $\alpha_{1\mathfrak{p}}, \ldots, \alpha_{n_\mathfrak{p}\mathfrak{p}}$ a basis of $E_\mathfrak{p}/E_\mathfrak{p}^p$. Here

$$n_\mathfrak{p} = \begin{cases} 1 & \text{if } \chi(\mathfrak{p}) \neq p, \\ (k_\mathfrak{p} : \mathbb{Q}_p) + \delta(k_\mathfrak{p}) & \text{if } \chi(\mathfrak{p}) = p. \end{cases}$$

For $\delta(k) = 1$, $\chi(\mathfrak{p}) = p$ let $\pi_\mathfrak{p} = \pi$, $\alpha_{1\mathfrak{p}} = \alpha_1, \ldots, \alpha_{n_\mathfrak{p}-1,\mathfrak{p}} = \alpha_0$, $\alpha_{n_\mathfrak{p}\mathfrak{p}} = \alpha_0$ be chosen in such a way that they satisfy the conditions of Lemmas 10.6 and 10.10, respectively. For $p = 2$ and an infinite prime $\mathfrak{p}$ we have $E_\mathfrak{p} = \mathbb{R}^\times$, and $\alpha_{1\mathfrak{p}} = -1$ generates $E_\mathfrak{p}/E_\mathfrak{p}^2$.

For $\alpha \in k^\times$ we denote the idele whose $\mathfrak{p}$-component is α and whose other components are 1 also by α. Let $\mathfrak{P}$ be a fixed prime divisor of $\mathfrak{p}$ in k_S. For primes $\mathfrak{p} \in S$ let $\sigma_\mathfrak{p}$ be an element in G_S with the following properties:

(i) $\sigma_\mathfrak{p}$ is a lift of the Frobenius automorphism of $\mathfrak{P}$ with respect to the maximal subextension of k_S/k in which $\mathfrak{P}$ is unramified;

(ii) the restriction of $\sigma_\mathfrak{p}$ to the maximal abelian subextension $k_S^{(2)}/k$ of k_S/k equals $(\pi_\mathfrak{p}, k_S^{(2)}/k)$.

For all primes $\mathfrak{p} \in S$ and $\nu = 1, \ldots, n_\mathfrak{p}$, let $\tau_{\nu\mathfrak{p}}$ denote an element of G_S such that

(i) $\tau_{\nu\mathfrak{p}}$ is an element of the inertia group $\mathcal{T}_\mathfrak{P}$ of $\mathfrak{P}$ in k_S/k;

(ii) the restriction of $\tau_{\nu\mathfrak{p}}$ to $k_S^{(2)}/k$ equals $(\alpha_{\nu\mathfrak{p}}, k_S^{(2)}/k)$.

Let ω_ν be a lift of $(\alpha_\nu, k_S^{(2)}/k)$ to G_S, $\nu = 1, \ldots, h$.

The automorphisms $\omega_1, \ldots, \omega_h, \tau_{1\mathfrak{p}}, \ldots, \tau_{n_\mathfrak{p}\mathfrak{p}}$, $\mathfrak{p} \in S$, form a system C of generators of G_S. In order to prove this we have to show 1) that each open subgroup H of G_S contains almost all elements from C and 2) that C generates the group G_S.

1. The fixed field of H is a finite subextension of k_S/k, hence has only finitely many ramified primes. For the other $\mathfrak{p} \in S$ we have $\tau_{1\mathfrak{p}}, \ldots, \tau_{n_\mathfrak{p}\mathfrak{p}} \in H$.

2. By Burnside's Basis Theorem 4.10 it is sufficient to prove that C mod $G_S^p[G_S, G_S]$ generates $G_S/G_S^p[G_S, G_S] \simeq J/U_S J^p k^\times$, i.e., we only have to show that

$$\{i_1, \ldots, i_h\} \cup \{\alpha_{1\mathfrak{p}}, \ldots, \alpha_{n_\mathfrak{p}\mathfrak{p}} \,|\, \mathfrak{p} \in S\}$$

is a system of generators of J mod $U_S J^p k^\times$. But this follows by (11.10) and (11.11) from the definition of the i_ν, $\alpha_{\nu\mathfrak{p}}$.

In general, C is not a minimal system of generators. Under the assumption $Б_S = 0$ it follows from (11.11) that we may omit

$$\dim Б_\varnothing = \dim\ E/p + \dim\ \mathrm{Cl}\,(k)[p]$$

elements from C.

To this end we determine a minimal system of generators of J mod $U_S J^p k^\times$. First assume that k is a number field.

By construction, $\{\alpha_{1\mathfrak{p}}, \ldots, \alpha_{n_\mathfrak{p}\mathfrak{p}} \,|\, \mathfrak{p} \in S\}$ is a minimal system of generators of U mod $U_S U^p$. We now have to omit elements in such a way that the remaining elements together with $\ker \varphi_3 = \mathrm{im}\ \varphi_2$ is a minimal system of generators of U mod $U_S U^p$.

Let q_ν denote the smallest positive integer such that $i_\nu^{q_\nu} \in U k^\times$, and let $\beta_\nu \in k$ be an element with $i_\nu^{q_\nu}(\beta_\nu) \in U$, $\nu = 1, \ldots, h$. Then

$$\{(\varepsilon_\mu), i_\nu^{q_\nu}(\beta_\nu) \,|\, \mu = 1, \ldots, r-1+\delta(k),\ \nu = 1, \ldots, h\} \bmod U_S U^p$$

is a basis of im φ_2. Consider the corresponding system of equations

$$\left.\begin{aligned} (\varepsilon_\mu) &= \textstyle\prod_{\mathfrak{p}\in S} \gamma_{\mathfrak{p}\mu} \alpha_{1\mathfrak{p}}^{a_{1\mathfrak{p}\mu}} \cdots \alpha_{n_\mathfrak{p}\mathfrak{p}}^{a_{n_\mathfrak{p}\mathfrak{p}\mu}}, && a_{\lambda\mathfrak{p}\mu} \in \mathbb{Z}_p, \\ & && \mu = 1, \ldots, r-1+\delta(k), \\ i_\nu^{a_\nu}(\beta_\nu) &= \textstyle\prod_{\mathfrak{p}\in S} \delta_{\mathfrak{p}\nu} \alpha_{1\mathfrak{p}}^{b_{1\mathfrak{p}\mu}} \cdots \alpha_{n_\mathfrak{p}\mathfrak{p}}^{b_{n_\mathfrak{p}\mathfrak{p}\mu}}, && b_{\lambda\mathfrak{p}\nu} \in \mathbb{Z}_p, \\ & && \nu = 1, \ldots, h, \end{aligned}\right\} \qquad (11.13)$$

where $\gamma_{\mathfrak{p}\mu}$ and $\delta_{\mathfrak{p}\nu}$ are products of a root of unity of order coprime to p and a principal unit from k if $\chi(\mathfrak{p}) \neq p$, a root of unity in k of order coprime to p if $\chi(\mathfrak{p}) = p$, and elements of $k_\mathfrak{p}^p$ if $\mathfrak{p}$ is an infinite prime. The coefficient matrix of this system is

$$\begin{pmatrix} a_{\lambda\mathfrak{p}\mu} \\ b_{\lambda\mathfrak{p}\nu} \end{pmatrix}_{\lambda_\mathfrak{p}\mathfrak{p};\mu\nu},$$

and its rank mod p is $r-1+\delta(k)+h$. By class field theory, (11.13) corresponds to the system of equations

$$\left.\begin{array}{l}\prod_{\mathfrak{p}\in S}\overline{\tau}_{1\mathfrak{p}}^{a_{1\mathfrak{p}\mu}}\cdots\overline{\tau}_{n_\mathfrak{p}\mathfrak{p}}^{a_{n_\mathfrak{p}\mathfrak{p}\mu}}=1,\quad \mu=1,\ldots,r-1+\delta(k),\\ \prod_{\mathfrak{p}\in S}\overline{\tau}_{1\mathfrak{p}}^{b_{1\mathfrak{p}\mu}}\cdots\overline{\tau}_{n_\mathfrak{p}\mathfrak{p}}^{b_{n_\mathfrak{p}\mathfrak{p}\mu}}=\overline{\omega}_\nu^{q_\nu},\ \nu=1,\ldots,h,\end{array}\right\} \tag{11.14}$$

from which we can determine which generators we can omit from C. A bar in (11.14) denotes the class mod $[G_S, G_S]$.

The considerations in the case where k is a function field are completely analogous. We only have to observe that there is no equation for i_1 since the divisor of i_1 has degree different from 0.

Let $C_0 \subseteq C$ be a minimal system of generators of G_S. The elements of $D = C \setminus C_0$ can be expressed with the help of (11.14) up to factors from $[G_S, G_S]$.

We now begin to determine a system of generating relations of G_S and first express $\sigma_\mathfrak{p}$ using elements of C_0.

Since $i_1, \ldots, i_h$ form a basis of J mod $UJ^pk^\times$, and since i_1 has minimal possible positive degree in the function field case, there is a $c_\mathfrak{p} \in \mathbb{Z}\setminus p\mathbb{Z}$ such that $\pi_\mathfrak{p}^{c_\mathfrak{p}}$ can be written as

$$\pi_\mathfrak{p}^{c_\mathfrak{p}} = \Big(\prod_{\nu=1}^{h} a_\nu^{c'_{\mathfrak{p}\nu}}\Big)\Big(\prod_{\mathfrak{q}} \alpha_\mathfrak{q}\alpha_{1\mathfrak{q}}^{c'_{1\mathfrak{q}\mathfrak{p}}}\cdots\alpha_{n_\mathfrak{q}\mathfrak{p}}^{c'_{n_\mathfrak{q}\mathfrak{q}\mathfrak{p}}}\Big)(\beta_\mathfrak{p}). \tag{11.15}$$

Here $\alpha_\mathfrak{q}$ is a product of a root of unity of order prime to p and a principal unit if $\chi(\mathfrak{q}) \neq p$, a principal unit of order coprime to p if $\chi(\mathfrak{q}) = p$, and $\alpha_\mathfrak{q} \in k_\mathfrak{q}^p$ if $\mathfrak{q}$ is archimedean. Moreover, $\beta_\mathfrak{p} \in k$, $c'_{\mathfrak{p}\nu}, c'_{\lambda_\mathfrak{q}\mathfrak{q}\mathfrak{p}} \in \mathbb{Z}_p$.

If we put

$$c_{\mathfrak{p}\nu} = c'_{\mathfrak{p}\nu}c'_\mathfrak{p}, \quad c_{\lambda_\mathfrak{q}\mathfrak{q}\mathfrak{p}} = c'_{\lambda_\mathfrak{q}\mathfrak{q}\mathfrak{p}}c_\mathfrak{p}^{-1},$$

then it follows from (11.15) that

$$\overline{\sigma}_\mathfrak{p} = \prod_{\nu=1}^{h}\overline{\omega}_\nu^{c_{\mathfrak{p}\nu}}\prod_{\mathfrak{q}\in S}\overline{\tau}_{1\mathfrak{q}}^{c_{1\mathfrak{q}\mathfrak{p}}}\cdots\overline{\tau}_{n_\mathfrak{q}\mathfrak{q}}^{c_{n_\mathfrak{q}\mathfrak{q}\mathfrak{p}}} \quad \text{for } \mathfrak{p}\in S. \tag{11.16}$$

Now let F be the free pro-p group with the system of generators

$$\{w_1, \ldots, w_h\} \cup \{t_{\lambda_\mathfrak{p}\mathfrak{p}} \mid \lambda_\mathfrak{p} = 1, \ldots, n_\mathfrak{p},\ \mathfrak{p}\in S,\ \tau_{\lambda_\mathfrak{p}\mathfrak{p}} \in C_0\}$$

corresponding to C_0, and

$$1 \longrightarrow R \longrightarrow F \xrightarrow{\ \psi\ } G_S \longrightarrow 1 \tag{11.17}$$

is the minimal presentation of G_S defined by

$$\begin{array}{l}\psi w_\nu = \omega_\nu, \quad \nu = 1, \ldots, h\\ \psi t_{\lambda_\mathfrak{p}\mathfrak{p}} = \tau_{\lambda_\mathfrak{p}\mathfrak{p}}, \quad \mathfrak{p}\in S,\ \lambda_\mathfrak{p} = 1, \ldots, n_\mathfrak{p},\ \tau_{\lambda_\mathfrak{p}\mathfrak{p}} \in C_0.\end{array}$$

For $\tau_{\lambda_\mathfrak{p}\mathfrak{p}} \in D$ let $t_{\lambda_\mathfrak{p}\mathfrak{p}}$ be a preimage in F that is given mod $[F, F]$ by the presentation of $\overline{\tau}_{\lambda_\mathfrak{p}\mathfrak{p}}$ by C_0 mod $[G, G]$ just as in (11.14). Next let $s_\mathfrak{p}$ be a preimage of $\sigma_\mathfrak{p}$ in F given mod $[F, F]$ by the presentation (11.16) of $\overline{\sigma}_\mathfrak{p}$, i.e.,

$$s_{\mathfrak{p}} = \prod_{\nu=1}^{h} w_{\nu}^{c_{\mathfrak{p}\nu}} \prod_{\mathfrak{q}\in S} t_{1\mathfrak{q}}^{c_{1\mathfrak{q}\mathfrak{p}}} \cdots t_{n_{\mathfrak{q}}\mathfrak{q}}^{c_{n_{\mathfrak{q}}\mathfrak{q}\mathfrak{p}}} s'_{\mathfrak{p}}, \quad s'_{\mathfrak{p}} \in [F, F]. \tag{11.18}$$

Since $\text{Б}_S = 0$, the map

$$\mathrm{H}^2(G_S) \longrightarrow \coprod_{\mathfrak{p}\in S} \mathrm{H}^2(G_{\mathfrak{p}})$$

is injective by Theorem 11.3, and if $\delta(k) = 1$, then by Theorem 11.4 even

$$\mathrm{H}^2(G_S) \longrightarrow \coprod_{\mathfrak{p}\in S\setminus\{\mathfrak{p}_0\}} \mathrm{H}^2(G_{\mathfrak{p}})$$

is injective, where $\mathfrak{p}_0$ denotes an arbitrary prime in S. We want to apply theorem 6.14 and use the results of Chapter 10.

Since $G_{\mathfrak{p}}$ gets mapped to the decomposition group $\mathcal{Z}_{\mathfrak{p}}$ by $\varphi_{\mathfrak{p}}$, we may choose the generators σ, τ_1, ..., $\tau_{n_{\mathfrak{p}}}$ of $G_{\mathfrak{p}}$ in such a way that

$$\varphi_{\mathfrak{p}}\sigma = \sigma_{\mathfrak{p}}, \quad \varphi_{\mathfrak{p}}\tau_1 = \tau_{1\mathfrak{p}}, \ldots, \varphi_{\mathfrak{p}}\tau_{n_{\mathfrak{p}}} = \tau_{n_{\mathfrak{p}}\mathfrak{p}}.$$

Thus, by Section 6.2, to every $\mathfrak{p} \in S$ there corresponds an exact and commutative diagram

$$\begin{array}{ccccccccc} 1 & \longrightarrow & R & \longrightarrow & F & \longrightarrow & G & \longrightarrow & 1 \\ & & \uparrow & & \uparrow & & \uparrow & & \\ 1 & \longrightarrow & R_{\mathfrak{p}} & \longrightarrow & F_{\mathfrak{p}} & \longrightarrow & G_{\mathfrak{p}} & \longrightarrow & 1 \end{array}$$

For a finite prime $\mathfrak{p}$ with $\delta(k_{\mathfrak{p}}) = 1$ we get the following relation in R:

$$r_{\mathfrak{p}} = t_{1\mathfrak{p}}^{N(\mathfrak{p})-1}[t_{1\mathfrak{p}}^{-1}, s_{\mathfrak{p}}^{-1}] \quad \text{for } \chi(\mathfrak{p}) \neq p \text{ (Satz 10.2)}$$

(in this case we also write $t_{\mathfrak{p}}$ instead of $t_{1\mathfrak{p}}$),

$$r_{\mathfrak{p}} = t_{0\mathfrak{p}}^{q-1} t_{1\mathfrak{p}}^{\widehat{q}} [t_{0\mathfrak{p}}, s_{\mathfrak{p}}][t_{1\mathfrak{p}}, t_{2\mathfrak{p}}] \cdots [t_{n_{\mathfrak{p}}-1,\mathfrak{p}}, t_{n_{\mathfrak{p}}-1,\mathfrak{p}}] r'_{\mathfrak{p}}$$

with $r'_{\mathfrak{p}} \in F^{(3,q)} \cap F^{(2)}$, $t_{0\mathfrak{p}} = t_{n_{\mathfrak{p}}\mathfrak{p}}$ for $\chi(\mathfrak{p}) = p$, $q \neq 2$ (Theorem 10.12).

For a real prime $\mathfrak{p}$, the maximal 2-extension of $k_{\mathfrak{p}} = \mathbb{R}$ is the field $\mathbb{C}$. The corresponding relation in $\mathbb{R}$ thus has the form

$$r_{\mathfrak{p}} = t_{n\mathfrak{p}}^2 = t_{\mathfrak{p}}^2.$$

We now apply Theorem 6.14 to (11.18) and get

Theorem 11.10. *Let S be a set of primes in k with $\text{Б}_S = 0$. Then the relation module R in the minimal presentation (11.17) of G_S is generated by the relations $r_{\mathfrak{p}}$ for $\mathfrak{p} \in S$, $\delta(k_{\mathfrak{p}}) = 1$. If $\delta(k) = 1$, we may omit any of these relations.*

By our class field theoretic calculation over k, the relations $r_{\mathfrak{p}}$ are determined modulo $[F, F]^p\big[[F, F], F\big]$.

Proof. Since we have determined $s_{\mathfrak{p}}$ and $t_{\lambda_{\mathfrak{p}}\mathfrak{p}}$ up to factors from $[F,F]$, the claim follows from the structure of the relations $r_{\mathfrak{p}}$, where we have to observe that

$$F^{(3,p)} \cap F^{(2)} = [F,F]^p\big[[F,F],F\big].$$

□

Things are especially simple if k is the field of rational numbers, a complex quadratic number field with class number not divisible by p, or a function field with characteristic and class number not divisible by p. In each of these cases we have

$$\text{Б}_{\varnothing} = \delta(k).$$

If $\text{Б}_S = 0$ and $\delta(k) = 0$, we therefore cannot omit one of the generators $\tau_{1\mathfrak{p}}, \ldots, \tau_{n_{\mathfrak{p}}\mathfrak{p}}$, $\mathfrak{p} \in S$, whereas for $\text{Б}_S = 0$ and $\delta(k) = 1$, we can omit exactly one of them, say $\tau_{\kappa\mathfrak{p}_0}$. The system of generators then becomes

$$\begin{array}{ll}
\{\tau_{1\mathfrak{p}}, \ldots, \tau_{n_{\mathfrak{p}}\mathfrak{p}} \,|\, \mathfrak{p} \in S\} & \text{for } \delta(k) = 0,\ \chi(k) = 0,\\
\{\omega_1\} \cup \{\tau_{1\mathfrak{p}} \,|\, \mathfrak{p} \in S\} & \text{for } \delta(k) = 0,\ 0 \neq \chi(k) \neq p,\\
\{\tau_{1\mathfrak{p}}, \ldots, \tau_{n_{\mathfrak{p}}\mathfrak{p}} \,|\, \mathfrak{p} \in S\} \setminus \{\tau_{\kappa\mathfrak{p}_0}\} & \text{for } \delta(k) = 1,\ \chi(k) = 0,\\
\{\omega_1\} \cup \{\tau_{1\mathfrak{p}} \,|\, \mathfrak{p} \in S\} \setminus \{\tau_{1\mathfrak{p}_0}\} & \text{for } \delta(k) = 1,\ 0 \neq \chi(k) \neq p.
\end{array}$$

In the last two cases we omit the relation corresponding to $\mathfrak{p}_0$. It is easily seen that, with the exception of the case $\delta(k) = 1$, $\chi(k) = 0$, the relations are determined up to factors in $F^{(3)}$ by our class field theoretical computation over k.

Example 11.11. Let $k = \mathbb{Q}$ and $p \neq 2$. Then $\text{Б}_S = 0$ for all S. For a prime $q \in S$ we can put

$$\alpha_{1q} = \alpha_q = \begin{cases} \text{primitive root} \mod q & \text{if } q \equiv 1 \pmod p,\\ 1+p & \text{if } q = p. \end{cases}$$

Then we can determine the $c_{1q'q} = c_{q'q}$, $q', q \in S$, from

$$\begin{aligned}
\frac{1}{q} &\equiv \alpha_q^{c_{q'q}} \pmod{q'} \quad \text{for } q' \neq p,\\
\frac{1}{q} &= (1+p)^{c_{pq}}.
\end{aligned}$$

For $q \neq p$, we get the relation

$$r_q = t_q^{q-1} \prod_{q' \in S} [t_q, t_{q'}]^{c_{q'q}} r'_q, \quad r'_q \in F^{(3)}.$$

Example 11.12. Let $k = \mathbb{Q}$ and $p = 2$. Then $\text{Б}_s = 0$ if S contains ∞, 2 or a prime $q \equiv 3 \pmod 4$. For $q \in S \setminus \{2\}$ we can put

$$\alpha_{1q} = \alpha_q = \begin{cases} -1 & \text{for } q = \infty, \\ \text{positive primitive root } \bmod q & \text{for } q \equiv 1 \bmod 2. \end{cases}$$

For $q = 2$, the basis $\{2, 5, -1\}$ of $\mathbb{Q}_2^\times/\mathbb{Q}_2^{\times 2}$ satisfies the conditions of Lemma 10.10; therefore we can put $\alpha_{12} = -1$, $\alpha_{22} = 5$. $V_\varphi/\mathbb{Q}^{\times 2}$ is generated by -1. From this, the following relation between the generators τ_{12}, τ_{22} and τ_q, $q \in S \setminus \{2\}$ results:

$$\tau_{12}\tau_\infty \prod_{1 \equiv 1\ (p)} \tau_q^{(q-1)/2} \equiv 1 \pmod{G_S^{(2)}}.$$

In order to get a minimal system of generators, we therefore have to take one of the elements

$$\tau_{12}, \tau_\infty, \tau_q \ (q \equiv 3 \bmod 4)$$

out of the system of generators.

Now we assume that $\infty \in S$, and we omit τ_∞ from the system of generators and r_∞ from the set of relations of G_S.

For $q \neq 2$ and $a', q \in S$ we find $c_{q'q}$ from

$$\frac{1}{q} \equiv \alpha_q^{c_{q'q}} \pmod{q'} \quad \text{for } q' \equiv 1 \pmod 2,$$
$$c_{\infty q} = 0.$$

Moreover, we find

$$\frac{1}{q} = (-1)^{c_{12q}} 5^{c_{22q}}.$$

We get the following relations.

For $q \equiv 1 \pmod 2$:

$$r_q = t_q^{q-1} \prod_{q' \in S} [t_q, t_{q'}]^{c_{q'q}} [t_q, t_{12}^{c_{12q}} t_{22}^{c_{22q}}] r_{q'}, \quad r_q' \in F^{(3)},$$

for $q = 2$:

$$r_2 = t_{12}^2 \prod_{q' \in S'} [t_{22}, t_q]^{c_{q'2}} r_2', \quad r_2' \in F^{(3,2)} \cap F^{(2)},$$

where we have put $S' = S \setminus \{2, \infty\}$.

Example 11.13. Let $k = k_0(x)$ be a function field of prime characteristic $q \neq p$ and $\delta(k_0) = 0$. Then $\text{Б}_S = 0$ for all S. For $\mathfrak{p} \in S$ let $\alpha_{1\mathfrak{p}} = \alpha_\mathfrak{p}$ be a primitive $(N(\mathfrak{p})-1)$-th root of unity in $k_0[x]/\varphi_\mathfrak{p}(x)$, where $\varphi_\mathfrak{p}(x)$ denotes the normalized uniformizer for $\mathfrak{p}$. For the degree valuation u we put $\varphi_u(x) = \frac{1}{x}$.

Let i_1 be the idele with entry $\frac{1}{x}$ at u and 1 elsewhere. We may put $\pi_{\mathfrak{p}} = \varphi_{\mathfrak{p}}(x)$. Then $c_{\mathfrak{p}} = c_{\mathfrak{p}1}$ equals the degree of $\varphi_{\mathfrak{p}}(x)$. Moreover, we find

$$\frac{1}{\varphi_{\mathfrak{p}}(x)} \equiv \alpha_{\mathfrak{p}'}^{c_{\mathfrak{p}'\mathfrak{p}}} \bmod \mathfrak{p}'$$

for $\mathfrak{p}', \mathfrak{p} \in S$, and the relation $r_{\mathfrak{p}}$ corresponding to $\mathfrak{p} \in S$ takes the form

$$r_{\mathfrak{p}} = t_{\mathfrak{p}}^{N(\mathfrak{p})-1}[t_{\mathfrak{p}}, w_1]^{c_{\mathfrak{p}}} \prod_{\mathfrak{p}' \in S} [t_{\mathfrak{p}}, t_{\mathfrak{p}'}]^{c_{\mathfrak{p}'\mathfrak{p}}} r_{\mathfrak{p}'}, \quad r_{\mathfrak{p}'} \in F^{(3)}.$$

Example 11.14. Let $k = k_0(x)$ be a function field of prime characteristic $q \neq p$, $\delta(k_0) = 1$. Let m be the cardinality of the field of constants k_0, and let p^s denote the largest p-power dividing $m-1$. We have $s \geq 1$ since $\delta(k_0) = 1$.

Next we have $\text{Б}_S = 0$ in the case where S contains a prime divisor $\mathfrak{p}_0$ with

$$N(\mathfrak{p}_0) \not\equiv 1 \pmod{p^{s+1}}. \tag{11.19}$$

Let ζ denote a primitive $(m-1)$-th root of unity in k_0, and $\alpha_{1\mathfrak{p}} = \alpha_{\mathfrak{p}}$ a primitive $(N(\mathfrak{p})-1)$-th root of unity in $k_0[x]/\varphi_{\mathfrak{p}}(x)$ with

$$\alpha_{\mathfrak{p}}^{\frac{N(\mathfrak{p})-1}{m-1}} = \zeta$$

for finite primes $\mathfrak{p} \in S$ and $\alpha_{1\mathfrak{p}} = \alpha_{\mathfrak{p}} = \zeta$ for the degree valuation. Then $V_\varphi/k^{\times p}$ is generated by ζ, and we get the following relation between the generators $\tau_{\mathfrak{p}}$, $\mathfrak{p} \in S$, of G_S:

$$\prod_{\mathfrak{p} \in S} \tau_{\mathfrak{p}}^{\frac{N(\mathfrak{p})-1}{m-1}} \equiv 1 \pmod{G_S^{(2)}}.$$

Thus we can omit the generator $\tau_{\mathfrak{p}_0}$.

The values of i_1, $c_{\mathfrak{p}}$ and $c_{\mathfrak{p}'\mathfrak{p}}$ are determined as in Example 11.13. We have to observe, however, that now primes of degree 1 may be ramified. For the degree valuation $\mathfrak{u}$ we get

$$\frac{x^{c_{\mathfrak{p}}}}{\varphi_{\mathfrak{p}}(x)} \equiv \zeta^{c_{\mathfrak{u}\mathfrak{p}}} \bmod \mathfrak{u},$$

hence $c_{\mathfrak{u}\mathfrak{p}} = 0$.

Since

$$t_{\mathfrak{p}_0} \equiv \prod \mathfrak{p} \in S \setminus \{\mathfrak{p}_0\} t_{\mathfrak{p}}^{-\frac{N(\mathfrak{p})-1}{N(\mathfrak{p}_0)-1}} \bmod F^{(2)},$$

the relation corresponding to $\mathfrak{p} \in S \setminus \{\mathfrak{p}_0\}$ takes the form

$$r_{\mathfrak{p}} = t_{\mathfrak{p}}^{N(\mathfrak{p})-1}[t_{\mathfrak{p}}, w_1]^{c_{\mathfrak{p}}} \prod_{\mathfrak{p}' \in S \setminus \{\mathfrak{p}_0\}} [t_{\mathfrak{p}}, t_{\mathfrak{p}'}]^{d_{\mathfrak{p}'\mathfrak{p}}} r_{\mathfrak{p}'}, \quad r_{\mathfrak{p}'} \in F^{(3)},$$

where

$$d_{\mathfrak{p}'\mathfrak{p}} = c_{\mathfrak{p}'\mathfrak{p}} - c_{\mathfrak{p}_0\mathfrak{p}} \frac{N(\mathfrak{p})-1}{N(\mathfrak{p}_0)-1}.$$

11.5 The Structure of G_S in Special Cases

In certain cases, the results that we achieved in the last two sections suffice for completely determining the structure of G_S. In this section, we consider some of these cases, and we restrict ourselves to number fields.

(i) Let $\text{Б}_S = 0$ and $G_S^{(3,p)} = G_S^{(4,p)}$. If S is finite, then so is $d(G_S)$, and then the last condition can be checked by an algorithm developed by A.I. Skopin [61]. It can be proved using induction that $G_S^{(n,p)} = G_S^{(n+1,p)}$ for all $n \geq 3$, and since

$$\bigcap_{n=1}^{\infty} G_S^{(n,p)} = \{1\},$$

this implies $G_S^{(3,p)} = 1$. Thus G_S is finite in this case. For more details, we refer to H. Koch [32, p. 56] and [34] as well as to the following example.

Example 11.15. Let $k = \mathbb{Q}, p \neq 2$, $S = \{p_1, p_2\}$,

$$\begin{aligned} p_\nu &\equiv 1 \pmod{p}, \quad p_\nu \not\equiv 1 \pmod{p^2}\ \nu = 1,2 \\ p_1 &\not\equiv x^p \pmod{p_2} \quad \text{for } x \in \mathbb{Z}. \end{aligned}$$

Then G_S is isomorphic to the nonabelian group with order p^3 and exponent p^2.

(ii) Now we consider the case where the elements $\sigma_\mathfrak{p}$ and $\tau_\mathfrak{p}$ in the decomposition groups of tamely ramified primes $\mathfrak{p}$ in k_S form a minimal system of generators of G_S. We assume in addition that S contains at least one prime dividing p.

We want to prove the following theorem:

Theorem 11.16. *Let k be a number field and S a set of primes of k satisfying the following conditions:*

1. *S contains a nonempty subset S_1 of prime divisors of p; if $\delta(k) = 0$, then $\delta(k_\mathfrak{p}) = 0$ for all $\mathfrak{p} \in S_1$, and if $\delta(k) = 1$, then S_1 consists of a single element.*

2. *We have the inequality*

$$d = \sum_{\mathfrak{p} \in S_1} n_\mathfrak{p} - r - h + 1 - \delta(k) \geq 0.$$

3. *$S_2 = S - S_1$ consists of*
   - *h primes $\mathfrak{p}_\nu$ with $N(\mathfrak{p}_\nu) \equiv 1 \pmod{p}$, $\nu = 1, \ldots, h$, that form a basis of $\mathrm{Cl}(k)/p$,*
   - *d more primes $\mathfrak{p}_\nu$ with $N(\mathfrak{p}_\nu) \equiv 1 \pmod{p}$, $\nu = 1$, $\nu = h+1, \ldots, h+d$,*

- *and, for* $p = 2$, *the real infinite places* $\mathfrak{p}_{h+d+1}, \dots, \mathfrak{p}_u$.

4. *We have* $\text{Б}_S = 0$.

5. $\prod_{\mathfrak{p}\in S_1} E_\mathfrak{p}/E_\mathfrak{p}^p$ *is generated by elements of the form* $\prod_{\mathfrak{p}\in S_1} \alpha E_\mathfrak{p}^p$ *with* $\alpha \in k$, $v_\mathfrak{q}(\alpha) \equiv 0 \pmod p$ *for all primes* $\mathfrak{q}$ *in* k *that are not in* S_2.

Then G_S *has the presentation*

$$1 \longrightarrow R \longrightarrow F \longrightarrow G_S \longrightarrow 1,$$

where F *is the free pro-p group with system of generators*

$$\{s_{\mathfrak{p}_\nu}, t_{\mathfrak{p}_\mu} \mid \nu = 1, \dots, h + d,\ \mu = 1, \dots u\}$$

and where R *is generated as a normal subgroup of* F *by the relations*

$$t_{\mathfrak{p}_\nu}^{N(\mathfrak{P}_\nu)-1}[t_{\mathfrak{p}_\nu}^{-1}, s_{\mathfrak{p}_\nu}^{-1}], \quad \nu = 1, \dots, h + d,$$

and

$$t_{\mathfrak{p}_\mu}^2, \quad \mu = 1, \dots, u.$$

Proof. We can put

$$\omega_\nu = \sigma_{\mathfrak{p}_\nu}, \quad \nu = 1, \dots, h.$$

By Section 11.4, the elements

$$\{\sigma_{\mathfrak{p}_\nu}, \tau_{\mathfrak{p}_\mu} \mid \nu = 1, \dots, h + d,\ \mu = 1, \dots, u\}$$

form a minimal system of generators of G_S if the determinant

$$D = \begin{vmatrix} a_{\lambda_\mathfrak{p}\mathfrak{p}\nu} \\ b_{\lambda_\mathfrak{p}\mathfrak{p}\mu} \\ c_{\lambda_\mathfrak{p}\mathfrak{p}\kappa} \end{vmatrix} \tag{11.20}$$

is $\not\equiv 0 \pmod p$; here D has row indices $\mathfrak{p} \in S$, $\mathfrak{p} \mid p$, $\lambda_\mathfrak{p} = 1, \dots, n_\mathfrak{p}$ and the column indices $\nu = 1, \dots, h$, $\mu = 1, \dots, r - 1 + \delta(k)$, $\kappa = h + 1, \dots, h + d$, and it is the determinant of the following system of equations:

$$\begin{gathered}
\sigma_{\mathfrak{p}_\nu}^{q_\nu} \prod_{\mathfrak{p}\in S_2} \overline{\tau}_\mathfrak{p}^{-a_{\mathfrak{p}\nu}} = \prod_{\mathfrak{p}\in S_1} \prod_{\lambda_\mathfrak{p}=1}^{n_\mathfrak{p}} \overline{\tau}_{\lambda_\mathfrak{p}\mathfrak{p}}^{a_{\lambda_\mathfrak{p}\mathfrak{p}\nu}}, \quad \nu = 1, \dots, h, \\
\prod_{\mathfrak{p}\in S_2} \overline{\tau}_\mathfrak{p}^{-b_{\mathfrak{p}\mu}} = \prod_{\mathfrak{p}\in S_1} \prod_{\lambda_\mathfrak{p}=1}^{n_\mathfrak{p}} \overline{\tau}_{\lambda_\mathfrak{p}\mathfrak{p}}^{b_{\lambda_\mathfrak{p}\mathfrak{p}\mu}}, \quad \mu = 1, \dots, r - 1 + \delta(k), \\
\overline{\sigma}_{\mathfrak{p}\lambda} \prod_{i=1}^{h} \overline{\sigma}_{\mathfrak{p}i}^{-c_{\lambda i}} \prod_{\mathfrak{p}\in S_2} \overline{\tau}_\mathfrak{p}^{-c_{\mathfrak{p}\lambda}} = \prod_{\mathfrak{p}\in S_1} \prod_{\lambda_\mathfrak{p}=1}^{n_\mathfrak{p}} \overline{\tau}_{\lambda_\mathfrak{p}\mathfrak{p}}^{c_{\lambda_\mathfrak{p}\mathfrak{p}\lambda}}, \\
c_{\lambda_\mathfrak{p}\mathfrak{p}\lambda} = c_{\lambda_\mathfrak{p}\mathfrak{p}\mathfrak{p}_\lambda}, \quad \lambda = h + 1, \dots, h + d.
\end{gathered}$$

In fact, in this case and only in this case can the elements to be left out of $\mathfrak{C}$ be chosen from the set

$$\{\tau_{\lambda_\mathfrak{p}\mathfrak{p}} \mid \lambda = 1, \dots, n_\mathfrak{p},\ \mathfrak{p} \in S,\ \mathfrak{p} \mid p\} \tag{11.21}$$

and the remaining elements of (11.21) be replaced by $\sigma_{\mathfrak{p}_\lambda} \mid \lambda = h+1, \ldots, h+d$. As is easily seen, the determinant D in (11.20) is $\not\equiv 0 \pmod p$ if condition 5 is satisfied. Thus Theorem 11.16 follows from Theorem 11.10 by observing the fact that we can omit the relation corresponding to $\mathfrak{p} \mid p$, $\mathfrak{p} \in S$ for $\delta(k) = 1$, and that there is no relation corresponding to $\mathfrak{p} \mid p$, $\mathfrak{p} \in S$ if $\delta(k) = 0$. □

Example 11.17. Let $k = \mathbb{Q}$, $p \neq 2$ and $S = \{p, q\}$, where $q \not\equiv 1 \pmod{p}^2$ is prime. Then G_S is the quotient group of the free pro-p group F with generators s_q, t_q modulo the normal subgroup generated by the relation $t_q^{q-1}[t_q^{-1}, s_q^{-1}]$.

Example 11.18. Let $k = \mathbb{Q}$, $p = 2$ and $S = \{2, q, \infty\}$, where $q \equiv \pm 3 \pmod 8$ is prime. Then G_S is the quotient group of the free pro-p group F with generators s_q, t_q, t_∞ modulo the normal subgroup generated by the relations $t_q^{q-1}[t_q^{-1}, s_q^{-1}]$ and t_∞^2.

Example 11.19. Let $k = \mathbb{Q}(\sqrt{-23})$, $p = 3$ and $S = \{\mathfrak{p}_1, \mathfrak{p}_2, \mathfrak{q}, p\}$, where $\mathfrak{p}_1$, $\mathfrak{p}_2$ are the primes above 3, $q \not\equiv 1 \bmod 9$ is a prime inert in $k\mathbb{Q}$, and $\mathfrak{p}$ is a prime ideal that is not principal (k has class number 3). Then G_S is the quotient group of the free pro-p group F with generators $s_q, t_q, s_\mathfrak{p}, t_\mathfrak{p}$ modulo the normal subgroup generated by the relations

$$t_q^{q-1}[t_q^{-1}, s_q^{-1}],\ t_\mathfrak{p}^{N(\mathfrak{p})-1}[t_\mathfrak{p}^{-1}, s_\mathfrak{p}^{-1}].$$

12. On p-Class Groups and p-Class Field Towers

Now we apply the results of Chapter 11 to the p-class group, i.e., the p-Sylow subgroup of the ideal class group, and to the p-class field tower, that is, the maximal unramified p-extension. We restrict ourselves to algebraic number fields.

The starting point of our investigations is the following: let k be an algebraic number field (a finite extension of $\mathbb{Q}$) and K/k a finite p-extension. Let S be the set of primes in k that ramify in K/k. Then the p-class field tower $K_\varnothing$ of K is contained in k_S. We want to study the p-class group and the p-class field tower of K using the arithmetic of k. Let $H_K = \mathrm{Gal}(k_S/K)$; then we have

Theorem 12.1. *For every $\mathfrak{p} \in S$, fix a prime ideal $\mathfrak{P}$ in k_S above $\mathfrak{p}$. Let $T_\mathfrak{p}$ denote the inertia subgroup of $\mathfrak{P}$ with respect to k_S/k. Then* $\mathrm{Gal}(K_\varnothing/k)$ *is isomorphic to the factor group of G_S modulo the normal subgroup generated by the groups $T_\mathfrak{p} \cap H_K$ for $\mathfrak{p} \in S$.*

Proof. By Section 8.2, $T_\mathfrak{p} \cap H_K$ is the inertia subgroup of $\mathfrak{P}$ with respect to k_S/K. The fixed field of the normal subgroup generated by the $T_\mathfrak{p} \cap H_K$, $\mathfrak{p} \in S$, is by Theorem 8.3 the maximal p-extension of K in which the $\mathfrak{P}$ and therefore the $\mathfrak{p}$ are unramified. □

In the following, we draw some conclusions from Theorem 12.1; they are, however, far from exhausting the possible applications of this theorem.

12.1 A Criterion for Class Number Prime to p

Let K/k be a finite abelian p-extension, which corresponds by class field theory to a subgroup X of $\mathrm{Cl}(K)$. We only consider extensions K/k satisfying the following conditions:

A. Let S be the set of primes ramified in K/k. Then $B_S = 0$.

B. Let $X_\mathfrak{p}$ and $Y_\mathfrak{p}$ denote the preimage of X under the natural maps $E_\mathfrak{p} \longrightarrow \mathrm{Cl}(K)$ and $k_\mathfrak{p}^\times \longrightarrow \mathrm{Cl}(k)$, and put $S_1 = \{\mathfrak{p} \,|\, \mathfrak{p} \in S,\ \mathfrak{p} \mid p\}$. Then the map

$$U_S U^p \prod_{\mathfrak{p} \in S_1} X_\mathfrak{p} / U_S U^p \longrightarrow J/U_S J^p k^\times \tag{12.1}$$

induced from (11.11) is injective.

Condition A guarantees that the considerations in Section 11.4 may be applied. Condition B means the following: If $\mathfrak{p} \mid p$ for some $\mathfrak{p} \in S$, we pick the minimal system of generators $\{\tau_{1,\mathfrak{p}}, \ldots, \tau_{w_\mathfrak{p},\mathfrak{p}}\}$ of $T_\mathfrak{p}$ as a normal subgroup of the decomposition group $Z_\mathfrak{p} = Z_\mathfrak{P}$ (this differs slightly from what we did in Section 11.4) in such a way that $\tau_{1,\mathfrak{p}}, \ldots, \tau_{v_\mathfrak{p},\mathfrak{p}}$ is a minimal system of generators of $T_\mathfrak{p} H_K$ mod H_K as a normal subgroup of $Z_\mathfrak{p} H_K / H_K$ and such that the generators $\tau_{v_\mathfrak{p}+1,\mathfrak{p}}, \ldots, \tau_{w_\mathfrak{p},\mathfrak{p}}$ lie in $T_\mathfrak{p} \cap H_K$, then the minimal system of generators C_0 of G can be chosen in such a way that C_0 contains the elements $\tau_{v_\mathfrak{p}+1,\mathfrak{p}}, \ldots, \tau_{w_\mathfrak{p},\mathfrak{p}}$ for $\mathfrak{p} \in S$, $\mathfrak{p} \mid p$.

In particular, condition B is always satisfied if S does not contain any prime dividing p.

Let $V(X)$ be the $\mathbb{F}_p$-vector space $J/U_S J^p k^\times \prod_{\mathfrak{p} \in S_1} X_\mathfrak{p}$ and $V(X) \wedge V(X)$ its exterior product. For $\alpha \in k_\mathfrak{p}$ let $\overline{\alpha}$ denote the image of α under the natural map $k_\mathfrak{p} \longrightarrow V(X)$. Then we have

Theorem 12.2. *Let $p \neq 2$ and assume that K/k is an extension satisfying conditions A and B. Then the class number of K is coprime to p if and only if*

1. *K contains the p-class field tower of k;*
2. *$V(X) \wedge V(X)$ is generated by elements of the form*

$$\overline{\alpha} \wedge \overline{\beta} \quad \textit{with } \alpha, \beta \in k_\mathfrak{p}, \mathfrak{p} \in S;$$

3. *There is no proper closed subgroup X' of X with finite index and such that $k_\mathfrak{p}^\times \cap X' = k_\mathfrak{p}^\times \cap X$ for all primes $\mathfrak{p}$ of k.*

Remark. A corresponding but more complicated result holds for $p = 2$. In the following proof, we allow $p = 2$ at first.

Proof. We begin with a presentation of G_S as a quotient group of F as in (11.17) and use it to derive a presentation of $\mathrm{Gal}(K_\emptyset/k)$ as a quotient of F. To this end we choose a system of generators of the groups $\mathcal{T}_\mathfrak{p} \cap H_K$. If $\mathfrak{p} \in S$ is infinite, then $\mathcal{T}_\mathfrak{p}$ has order at most 2, and since $\mathcal{T} \cap H_K$ is a proper subgroup of $\mathcal{T}_\mathfrak{p}$, we have $\mathcal{T}_\mathfrak{p} \cap H_K = \{1\}$. For finite primes $\mathfrak{p} \in S$ that are tamely ramified, $\mathcal{T}_\mathfrak{p}$ is cyclic, and $\mathcal{T}_\mathfrak{p} \cap H_K$ is generated by a power $\tau_\mathfrak{p}^{q_\mathfrak{p}}$ of the generator $\tau_\mathfrak{p}$ of $\mathcal{T}_\mathfrak{p}$, where $p \mid q_\mathfrak{p} \mid N(\mathfrak{p}) - 1$. For $\mathfrak{p} \mid p$, $\mathfrak{p} \in S$, we choose a system of generators of $\mathcal{T}_\mathfrak{p}$ as a normal subgroup of $\mathcal{Z}_\mathfrak{p}$ as described above. Then $\mathcal{T}_\mathfrak{p} \cap H_K$ is generated as a normal subgroup of $\mathcal{Z}_\mathfrak{p}$ by

$$\tau_{1\mathfrak{p}}^{q_{1,\mathfrak{p}}}, \ldots, \tau_{v_\mathfrak{p}}^{a_{v_\mathfrak{p}\mathfrak{p}}}, \tau_{v_\mathfrak{p}+1}, \ldots, \tau_{w_\mathfrak{p}\mathfrak{p}} \quad \text{with } p \mid q_{\nu_\mathfrak{p}\mathfrak{p}},\ \nu_\mathfrak{p} = 1, \ldots, v_\mathfrak{p},$$

and the commutators $[\xi_\mathfrak{p}, \eta_\mathfrak{p}]$, where $\xi_\mathfrak{p}, \eta_\mathfrak{p} \in \mathcal{Z}_\mathfrak{p}$. Because of condition B we can choose a minimal system $\mathfrak{E}_0$ of generators of G_S in such a way that, for $\mathfrak{p} \mid p$, $\mathfrak{p} in S$, the elements $\tau_{v_\mathfrak{p}+1,\mathfrak{p}}, \ldots, \tau_{w_\mathfrak{p}\mathfrak{p}}$ are in $\mathfrak{E}_0$.

By Theorem 12.1, Gal $(K_\varnothing/k)$ is a quotient group of F: as in Section 11.4, we choose preimages $s_\mathfrak{p}$, $t_\mathfrak{p} = t_{1\mathfrak{p}}$ and $t_{\nu\mathfrak{p}}$ of $\sigma_\mathfrak{p}$, $\tau_\mathfrak{p}$ and $\tau_{\nu\mathfrak{p}}$, respectively, and form the quotient of F by the normal subgroup R_1 generated by R and the elements

$$t_\mathfrak{p}^{q_\mathfrak{p}} \quad \text{for} \quad N(\mathfrak{p}) \equiv 1 \pmod p, \ \mathfrak{p} \in S,$$
$$t_{1\mathfrak{p}}^{q_{1\mathfrak{p}}}, \ldots, t_{v_\mathfrak{p}\mathfrak{p}}^{q_{v_\mathfrak{p}\mathfrak{p}}}, t_{v_\mathfrak{p}+1,\mathfrak{p}}, \ldots, t_{w_\mathfrak{p}\mathfrak{p}}, \ \text{for } \mathfrak{p} \mid p, \ \mathfrak{p} \in S,$$
$$[x_\mathfrak{p}, y_\mathfrak{p}] \quad \text{with } x_\mathfrak{p}, y_\mathfrak{p} \in H_\mathfrak{p} \text{ for } \mathfrak{p} \mid p, \ \mathfrak{p} \in S,$$

where $H_\mathfrak{p}$ denotes the subgroup of F generated by $s_\mathfrak{p}$, $t_{1\mathfrak{p}}, \ldots, t_{w_\mathfrak{p}\mathfrak{p}}$.

Observing the structure of $\rho_\mathfrak{p}$ we find that R_1 is generated by the elements

$$\begin{array}{lll} t_\mathfrak{p}^{q_\mathfrak{p}}, [x_\mathfrak{p}, y_\mathfrak{p}] \text{ with } x_\mathfrak{p}, y_\mathfrak{p} \in H_\mathfrak{p} & \text{for } N(\mathfrak{p}) \equiv 1 \pmod p, & \mathfrak{p} \in S, \\ t_\mathfrak{p}^2 & \text{for } \mathfrak{p} \text{ real}, & \mathfrak{p} \in S, \\ t_{1\mathfrak{p}}^{q_{1\mathfrak{p}}}, \ldots, t_{v_\mathfrak{p}\mathfrak{p}}^{q_{v_\mathfrak{p}\mathfrak{p}}}, t_{v_\mathfrak{p}+1,\mathfrak{p}}, \ldots, t_{w_\mathfrak{p}\mathfrak{p}}, & [x_\mathfrak{p}, y_\mathfrak{p}] \text{ with } x_\mathfrak{p}, y_\mathfrak{p} \in H_\mathfrak{p} & \\ & \text{for } \mathfrak{p} \mid p, & \mathfrak{p} \in S. \end{array}$$

The class number of K is not divisible by p if and only if Gal $(K_\varnothing/K)$ is abelian, and so X is the subgroup of C_k corresponding to $K_\varnothing$ by class field theory. Gal $(K_\varnothing/K)$ is abelian if and only if $[F, F] \subseteq R_1$.

We now assume that $p \neq 2$. If $a_1, \ldots, a_m$ are arbitrary elements of F, then (7.7) shows that $a_1^{p\nu_1} \cdots a_m^{p\nu_m} \in F^{(2)}$ already implies

$$a_1^{p\nu_1} \cdots a_m^{p\nu_m} \in F^{(2)} F^{(3)}.$$

Thus the condition $R_1 \supseteq [F, F]$ is equivalent to the fact that $F^{(2)}$ is generated as a normal subgroup of F by elements of the form $[x_\mathfrak{p}, y_\mathfrak{p}]$ with $x_\mathfrak{p}, y_\mathfrak{p} \in H_\mathfrak{p}$, $\mathfrak{p} \in S$, $\mathfrak{p}$ finite, and $[x, t_{\nu_\mathfrak{p}\mathfrak{p}}]$ with $x \in F$, $\nu_\mathfrak{p} = v_\mathfrak{p} + 1, \ldots, w_\mathfrak{p}$, $\mathfrak{p} \in S$, $\mathfrak{p} \mid p$.

Let N denote the normal subgroup of F generated by the $t_{\nu_\mathfrak{p}\mathfrak{p}}$, and let M denote the normal subgroup of G_S generated by the $\tau_{\nu_\mathfrak{p}\mathfrak{p}}$, where $\nu_\mathfrak{p} = v_\mathfrak{p} + 1, \ldots, w_\mathfrak{p}$, $\mathfrak{p} \in S$, $\mathfrak{p} \mid p$. By condition B we have

$$F/F^{(2,p)}N \simeq G_S/G_S^{(2,p)}M \simeq J/U_S J^p k^\times \prod_{\mathfrak{p} \in S_1} X_\mathfrak{p} = V(X).$$

Now for arbitrary finitely generated free pro-p groups F_1, the $\mathbb{F}_p$-vector space $F_1^{(2)}/F_1^{(2)}F_1^{(3)}$ is isomorphic to the exterior product of $F_1/F_1^{(2,p)}$ (see M. Hall [21, Ch. 11.1]); this implies the claim. □

We now want to give examples in which the conditions A and B are satisfied. To this end, we first prove the following theorem that will also be important in Chapter 13.

Theorem 12.3. *If $\delta(k) = 1$ and if S is a set of primes in k containing all primes dividing p, then $\mathbf{Б}_S$ is isomorphic to the quotient group $\mathrm{Cl}_S(k)/p$ of $\mathrm{Cl}\,(k)/p$ modulo the subgroup of classes generated by primes $\mathfrak{p} \in S$.*

Proof. Let $\overline{k}$ be the compositum of all unramified cyclic extensions of degree p in which the primes of S split completely. By Kummer's Theory of cyclic extensions of degree p over k, the map

$$\sqrt[p]{\alpha} \longmapsto \chi(\alpha)\sqrt[p]{\alpha}, \quad \alpha \in V_S$$

assigns to every $\chi \in \mathbf{Б}_S$ an element of the Galois group $\mathrm{Gal}\,(\overline{k}/k)$. This map induces an isomorphism $\mathbf{Б}_S \longrightarrow \mathrm{Gal}\,(\overline{k}/k)$. On the other hand, class field theory (see Section 8.8) shows that $\mathrm{Gal}\,(\overline{k}/k) \simeq \mathrm{Cl}_S(k)/p$. □

Example 12.4. Let ζ_{p^n} be a primitive p^n-th root of unity. Then the class number of $\mathbb{Q}(\zeta_{p^n})$ is divisible by p if and only if the class number of $\mathbb{Q}(\zeta_p)$ is.

The claim of Example 12.4 can be derived without problems directly from Theorem 12.1: Let $\mathfrak{p}_n$ be the unique prime ideal dividing p in $k_n = \mathbb{Q}(\zeta_{p^n})$ and put $S = \{\mathfrak{p}_1\}$. If the class number of k_1 is coprime to p, then $\mathfrak{p}_1$ ramifies completely in k_S/k_1; thus $\mathfrak{p}_n$ is ramified completely in k_S/k_n, and by Theorem 12.1, the class number of k_n is coprime to p. The converse follows directly from class field theory.

Example 12.5. Let $p \neq 2$ and $k = \mathbb{Q}$. If there are more than three primes ramified in $K/\mathbb{Q}$, then the class number of K is divisible by p.

12.2 The p-Class Field Tower of Cyclic Extensions of Degree p

We now specialize the considerations in Section 12.1 by assuming in addition to conditions A and B that the base field k has class number coprime to p and that the extension K/k is cyclic of degree p. We want to study generators and relations of $\mathrm{Gal}\,(K_\varnothing/K)$.

We continue to use the notation from Section 12.1. Because of our additional assumptions, several things simplify:

- Since $\mathrm{Cl}\,(k)/p = 0$, the minimal system of generators $\mathfrak{E}_0$ contains only automorphisms from the groups $\mathcal{T}_\mathfrak{p}$, $\mathfrak{p} \in S$. Since K/k is cyclic of degree p, we get

$$v_\mathfrak{p} = 1 \quad \text{for } \mathfrak{p} \mid p, \mathfrak{p} \in S,$$

and

$$q_\mathfrak{p} = q_{1\mathfrak{p}} = p \quad \text{for } \mathfrak{p} \in S.$$

- The module R_1 of relations is generated by

$$\begin{array}{lll} t_{\mathfrak{p}}^p, [t_{\mathfrak{p}}, s_{\mathfrak{p}}] & \text{for } N(\mathfrak{p}) \equiv 1 \pmod p, & \mathfrak{p} \in S, \\ t_{\mathfrak{p}}^2 & \text{for } \mathfrak{p} \text{ real}, & \mathfrak{p} \in S, \\ t_{\mathfrak{p}}^p, [t_{\mathfrak{p}}, s_{\mathfrak{p}}], t_{2\mathfrak{p}}, \ldots, t_{w_{\mathfrak{p}}\mathfrak{p}} & \text{for } \mathfrak{p} \mid p, & \mathfrak{p} \in S. \end{array}$$

  Here we have put $t_{1\mathfrak{p}} = t_{\mathfrak{p}}$.

In order to be able to treat all finite primes in the same way, we form the quotient of F modulo the normal subgroup N generated by the elements $t_{2\mathfrak{p}}, \ldots, t_{w_{\mathfrak{p}}\mathfrak{p}}$ for all $\mathfrak{p} \mid p$, $\mathfrak{p} \in S$, and set

$$F_1 = F/N, \quad R_2 = R_1/N.$$

Let u denote the number of primes in S, and n the number of primes $\mathfrak{p}$ with $\tau_{\mathfrak{p}} \in C_0$. We now write $S = \{\mathfrak{p}_1, \ldots, \mathfrak{p}_u\}$ and put

$$t_\nu = t_{\mathfrak{p}_\nu} N, \; s_\nu = s_{\mathfrak{p}_\nu} N, \quad \nu = 1, \ldots, u.$$

Then F_1 is a free pro-p group with minimal system of generators $t_1, \ldots, t_n$, and R_2 is generated as a normal subgroup of F_1 by the elements

$$t_1^p, \ldots, t_u^p, \; [t_1, s_1], \ldots, [t_u, s_u],$$

where we have put

$$s_\nu = 1$$

for infinite primes $\mathfrak{p}_\nu$.

We want to prove the following theorem:

Theorem 12.6. *Let K/k be a cyclic extension of degree p satisfying conditions A and B. Assume that the class number of k is not divisible by p. Moreover, let $m(X)$ denote the dimension of the subspace of $V(X)$ generated by the elements $\overline{\alpha}_{\mathfrak{p}}$, where $\alpha_{\mathfrak{p}} \in Y_{\mathfrak{p}}$, $\mathfrak{p} \in S$.*

Then the generator rank $d(\varnothing)$ of $\mathrm{Gal}(K_\varnothing/K)$ satisfies

$$2(n-1) - m(X) \le d(\varnothing) \le (p-1)(n-1) - (p-2)m(X)$$

if $p \neq 2$, and

$$d(\varnothing) = n - 1$$

if $p = 2$.

Remark. For $p = 3$ we get

$$d(\varnothing) = 2(n-1) - m(X).$$

Proof of Theorem 12.6. Let H denote the preimage of $\mathrm{Gal}(K_\varnothing/K)$ under the map $F_1 \longrightarrow \mathrm{Gal}(K_\varnothing/K)$. Then H is a free pro-p group of index p in F_1, and we have the (non-minimal) presentation

$$1 \longrightarrow R_2 \longrightarrow H \longrightarrow \mathrm{Gal}(K_\varnothing/K) \longrightarrow 1.$$

Since all the $\mathfrak{p} \in S$ are ramified in K/k, the element t_ν is not in H for any $\nu = 1, \ldots, u$. Thus we may assume without loss of generality that $t_1, \ldots, t_u$ are chosen in such a way that $t_1 t_\nu^{-1}$ and hence $t_\mu t_\nu^{-1}$ is in H for $\nu, \mu = 1, \ldots, u$. Then it is easily seen that H is generated as a pro-p group by the elements

$$t_1^p,\, t_1^\nu t_1 t_\mu^{-1} t_1^{-\nu} = v_{\mu\nu} \quad \text{with } \nu = 0, \ldots, p-1,\ \mu = 2, \ldots, n. \tag{12.2}$$

There are exactly $1 + p(n-1)$ elements in (12.2), so Example 6.3 shows that it is a minimal system of generators of H.

Since

$$t_1^p = v_{\mu,p-1} v_{\mu,p-2} \cdots v_{\mu,0} t_\mu^p, \tag{12.3}$$

(12.2) may be replaced by the system of generators

$$\{t_1^p, t_\mu^p, v_{\mu\nu} \mid \nu = 0, \ldots, p-2,\ \mu = 2, \ldots, n\}.$$

Let Q be the normal subgroup of H generated by the elements $t_1^p, \ldots, t_n^p$. Then Q is even normal in F_1; in fact, we have

$$t_1 t_\mu^p t_1^{-1} = t_1 t_\mu^{-1} t_\mu^p (t_1 t_\mu^{-1})^{-1} \in Q.$$

Since $t_\mu^p \in R_2$ for $\mu = 1, \ldots, n$, we may go over to the quotient groups $F_2 = F_1/Q$, $H_1 = H/Q$. $R_3 = R_2/Q$ and get the presentation

$$1 \longrightarrow R_3 \longrightarrow H_1 \longrightarrow \mathrm{Gal}(k_\varnothing/K) \longrightarrow 1, \tag{12.4}$$

where H_1 is a free pro-p group with minimal system of generators

$$\{\overline{v}_{\mu\nu} \mid \nu = 0, \ldots, p-2,\ \mu = 2, \ldots, n\}. \tag{12.5}$$

For $p = 2$, the presentation (12.4) is minimal, since $\{t_1 t_\mu^{-1} \mid \mu = 2, \ldots, n\}$ together with t_1 forms a minimal system of generators of F_1.

For $p \geq 3$, the presentation (12.4) is in general not minimal. In order to prove the estimate in Theorem 12.6, we have to consider the structure of the relations in R_3.

The relations of the form $\overline{t}_\mu^p$, $\mu = n+1, \ldots, u$ are in $H_1^{(2,p)}$, as the following lemma shows.

Lemma 12.7. *If $g \in F_1$, then $g^p \in QH^{(2,p)}$.*

Proof. For $g \in H$ there is nothing to prove. Assume therefore that $g \notin H$. Then g satisfies the congruence

$$g \equiv t_1^\nu \prod_{\mu=2}^{n} \prod_{\kappa=0}^{p-1} v_{\mu\kappa}^{a_{\mu\kappa}} \pmod{QH^{(2,p)}}$$

for certain ν, $a_{\mu\kappa} \in \mathbb{Z}$ with $0 \le \nu \le p-1$. In congruences mod Q, the index κ in $v_{\mu\kappa}$ can be reduced mod p. From

$$v_{\mu\kappa} t_1^\nu \equiv t_1^\nu v_{\mu,\kappa-\nu} \pmod{Q},$$

we get

$$g^p \equiv t_1^{p\nu} \prod_{\mu=2}^{n} \prod_{\kappa=0}^{p-1} v_{\mu\kappa}^{b_{\mu\kappa}} \pmod{QH^{(2,p)}}$$

with

$$b_{\mu\kappa} = \sum_{i=0}^{p-1} a_{\mu,\kappa+\nu i},$$

hence

$$b_{\mu\kappa} = b_{\mu 0} \quad \text{for } \kappa = 1, \ldots, p-1.$$

By (12.3), this implies the claim. □

Lemma 12.8. *The elements*

$$w_\mu = \overline{v}_{\mu 0},\ w_{\mu 1} = [\overline{v}_{\mu 0}, \overline{t}_1], \ldots, w_{\mu,p-2} = [\overline{v}_{\mu 0}, \overline{t}_1, \ldots, \overline{t}_1],\ \mu = 2, \ldots, n,$$

form a minimal system of generators of H_1.

Proof. The system of generators (12.5) is clearly equivalent to the system

$$\{\overline{v}_{\mu 0}, [\overline{v}_{\mu 0}, \overline{t}_1^\nu] \mid \mu = 2, \ldots, n, \nu = 1, \ldots, p-2\}.$$

Since

$$[x, t_1^\kappa] = [x, t_1^{\kappa-1}][x, t_1][x, t_1, t_1^{\kappa-1}] \quad \text{for } x \in H,$$

this implies the claim by induction on the weight of $[\overline{v}_{\mu 0}, \overline{t}_1, \ldots, \overline{t}_1]$. □

Now, as in (11.18), let

$$s_\nu \equiv t_1^{c_{1\nu}} t_2^{c_{2\nu}} \cdots t_n^{c_{n\nu}} \pmod{QF_1^{(2)}}.$$

We may assume without loss of generality that $s_\nu \in H$ for $\nu = 1, \ldots, u$. Then

$$s_\nu \equiv v_{20}^{-c_{2\nu}} \cdots v_{n0}^{-c_{n\nu}} \pmod{QF_1^{(2)}}$$

and

$$[s_\nu, t_1] \equiv [v_{20}, t_1]^{-c_{2\nu}} \cdots [v_{n0}, t_1]^{-c_{n\nu}} \pmod{QF_1^{(3)}}.$$

If the elements

$$\overline{[v_{20}, t_1]}^{-c_{2\nu}} \cdots \overline{[v_{n0}, t_1]}^{-c_{n\nu}} H_1^{(2,p)} F_2^{(3,p)}, \quad \nu = 1, \dots, u,$$

in $H_1/H_1^{(2,p)} F_2^{(3,p)}$ generate a subspace of dimension m, then the elements

$$\overline{[s_\nu, t_1]} H_1^{(2,p)}, \quad \nu = 1, \dots, u,$$

and the elements

$$\overline{[s_\nu, t_1, t_1]} H_1^{(2,p)}, \dots, \overline{[s_\nu, \underbrace{t_1, \dots, t_1}_{p-2}]} H_1^{(2,p)}, \quad \nu = 1, \dots, u,$$

derived from them generate a subspace of $H_1/H_1^{(2,p)}$ whose dimension is at least

$$(p-2)m$$

and at most

$$m + (n-1)(p-3).$$

This gives the bounds

$$2(n-1) - m \le d(\varnothing) \le (p-1)(n-1) - (p-2)m$$

for $d(\varnothing)$. For the proof of Theorem 12.6, it is therefore enough to show that $m = m(X)$.

To this end we construct a monomorphism

$$\varphi : X/J^p U_S k^\times \textstyle\prod_{\mathfrak{p} \in S_1} X_\mathfrak{p} \xrightarrow{\sim} H/F_1^{(2,p)} \longrightarrow H_1/H_1^{(2,p)} F_2^{(3,p)}.$$

For $g \in H$, we put

$$\varphi(g F_1^{(2,p)}) = \overline{[g, t_1]} H_1^{(2,p)} F_2^{(3,p)}.$$

It is easy to see that φ is a well defined homomorphism. Now let $g F_1^{(2,p)} \in \ker \varphi$ and

$$g \equiv v_{20}^{a_2} \cdots v_{n0}^{a_n} \pmod{F_1^{(2,p)}}$$

with $0 \le a_\nu \le p-1$ for $\nu = 2, \dots, n$. Then

$$1 = \varphi(g F_1^{(2,p)}) = w_{21}^{a_2} \cdots w_{n1}^{a_n} H_1^{(2,p)} F_2^{(3,p)},$$

and now Lemma 12.8 and Theorem 7.22 imply

$$a_2 = a_3 = \ldots = a_n = 0.$$

Now the claim follows from the injectivity of φ. □

We now study the relation module R_3 more closely.

Theorem 12.9. *Assume that the conditions in Theorem 12.6 are satisfied. Then* $\mathrm{Gal}(K_\varnothing/K)$ *has a presentation*

$$1 \longrightarrow R_3 \longrightarrow H_1 \longrightarrow \mathrm{Gal}(K_\varnothing/K) \longrightarrow 1, \tag{12.6}$$

where H_1 *is a free pro-p group with minimal system of generators*

$$\{t_1^{\nu+1}t_\mu^{-1}t_1^{-\nu}Q \mid \nu = 0,\dots,p-1,\ \mu = 2,\dots,n\}.$$

R_3 *is generated as a normal subgroup of* H_1 *by the system of relations*

$$\{t_\mu^p Q, [s_\nu, t_\nu^\kappa]Q \mid \mu = n+1,\dots,u,\ \nu = 1,\dots,u,\ \kappa = 1,\dots,p-1\}. \tag{12.7}$$

The presentation (12.6) is minimal for $p = 2$.

Proof. By the preceding, it only remains to show that the system (12.7) of relations generates R_3 as a normal subgroup of H_1.

By induction on κ it is first shown that

$$[s_\nu, t_\nu^\kappa] \in R_s.$$

Here the identity

$$[s_\nu, t_\nu^{\kappa+1}] = [s_\nu, t_\nu][s_\nu, t_\nu^\kappa][s_\nu, t_\nu^\kappa, t_\nu]$$

has to be used.

The system (12.7) of relations generates R_3 as a normal subgroup of F. Moreover, we have

$$\begin{aligned} t_1 t_\mu^p t_1^{-1} &\equiv t_1 t_\mu^{-1} t_\mu^p (t_1 t_\mu^{-1})^{-1}, \\ t_1^{-1}[s_\nu, t_\nu^\kappa]t_1 &= t_1^{-1}t_\nu t_\nu^{-1}[s_\nu, t_\nu^\kappa]t_\nu(t_1^{-1}t_\nu)^{-1}, \\ t_\mu^{-1}[s_\nu, t_\nu^\kappa]t_\mu &= [s_\nu, t_\nu^\kappa][s_\nu, t_\nu^\kappa, t_\nu] \ = \ [s_\nu, t_\nu]^{-1}[s_\nu, t_\nu^{\kappa+1}]. \end{aligned}$$

Since

$$[s_\nu, t_\nu^\kappa] \equiv [s_\nu, t_\nu^{\kappa+p}] \pmod{Q},$$

this implies the claim. □

The following theorem shows that the group $\mathrm{Gal}(K_\varnothing/K)$ can be computed $\mathrm{mod}\,\mathrm{Gal}(K_\varnothing/K)^{(3,p)}$ using class field theory over k for $p = 2$ and in special cases for $p \geq 3$.

Theorem 12.10. *Let the assumptions be the same as in Theorem 12.6; for* $p \geq 3$, *we assume in addition that* $m(X) = n - 1$. *Then*

$$F_1^{(\kappa,p)} R_2 = H^{(\kappa,p)} R_2 \quad \textit{for } \kappa = 2, 3, \dots.$$

Proof. We first remark that $H^{(\kappa,p)}$ is a normal subgroup of F_1 for all $\kappa = 1, 2, \ldots$. We prove Theorem 12.10 by induction on κ.

Since $m(X) = n - 1$, we have

$$[v_{\mu 0}, t_1] \in H^{(2,p)} R_2, \quad \mu = 2, \ldots n \tag{12.8}$$

for $p \geq 3$. Equation (12.8) also holds for $p = 2$, as is immediately checked directly (more exactly, we have $[v_{\mu 0}, t_1] \in H^{(2,2)} Q$).

Assume that, for $g_0, g_1, g_2 \in F$ we already have shown that

$$[g_0, g_1],\ [g_0, g_2] \in H^{(2,p)} R_2.$$

Then

$$[g_0, g_1 g_2] = [g_0, g_2][g_0, g_1][g_0, g_1, g_2] \in H^{(2,p)} R_2. \tag{12.9}$$

Thus it follows from (12.8) that

$$[v_{\mu 0}, g] \in H^{(2,p)} R_2 \tag{12.10}$$

for all $g \in F_1$. From (12.10) we get, using (12.9),

$$[h, g] \in H^{(2,p)} R_2$$

for $h \in H$, $g \in F_1$, and since

$$[t_1, t_1^{\nu} h] = [t_1, h] = [h, t_1]^{-1},$$

we have

$$[g_1, g_2] \in H^{(2,p)} R_2 \tag{12.11}$$

for all $g_1, g_2 \in F_1$.

Moreover, we get

$$(t_1^{\nu} h)^p \equiv t_1^{\mu p} h^p \equiv 1 \pmod{H^{(2,p)} R_2}. \tag{12.12}$$

Together with (12.11), (12.12) proves the claim in the case $\kappa = 2$.

We now assume that we have proved

$$F_1^{(\kappa,p)} R_2 = H^{(\kappa,p)} R_2 \tag{12.13}$$

for some $\kappa \geq 2$, and we claim

$$F_1^{(\kappa+1,p)} R_2 = H^{(\kappa+1,p)} R_2. \tag{12.14}$$

For $g_\kappa \in F^{(\kappa,p)}$, $g \in F$ we have to show that

$$[g_\kappa, g], g_\kappa^p \in H^{(\kappa+1,p)} R_2. \tag{12.15}$$

To this end we use the identity

$$y^{-1}[x,y^{-1},z]y \cdot z^{-1}[y,z^{-1},x]z \cdot x^{-1}[z,x^{-1},y]x = 1$$

for $x = h_{\kappa-1}$, $y = h^{-1}$, $z = g$. Now $[g, h_{\kappa+1}^{-1}, h^{-1}] \in H^{(\kappa+1,p)}R_2$ by (12.13), and $[h^{-1}, g^{-1}, h_{\kappa-1}] \in H^{(\kappa+1,p)}R_2$ by (12.11) and (7.4), so we deduce that

$$[h_{\kappa-1}, h, g] \in H^{(\kappa+1,p)}R_2.$$

Moreover, we have

$$[h_{\kappa-1}^p, g] \equiv [h_{\kappa-1}, g]^p \equiv 1 \pmod{H^{(\kappa+1,p)}R_2}.$$

□

Remark. Let $G = \mathrm{Gal}(K_\varnothing/K)$. By Theorem 12.10 we have

$$G/G^{(3,p)} = H/H^{(3,p)}R_2 = H/F_1^{(3,p)}R_2.$$

For computing $G/G^{(3,p)}$ it is therefore sufficient to know the relations R_2 mod $F_1^{(3,p)}$.

Theorem 12.11 *Let the assumptions be the same as in Theorem 12.6. For $p = 2$ we have*

$$F_1^{(\kappa,p)}Q = H^{(\kappa,p)}Q, \quad \kappa = 2, 3, \ldots,$$

and therefore

$$F_2^{(\kappa,p)} = H_1^{(\kappa,p)}, \quad \kappa = 2, 3, \ldots.$$

The proof of Theorem 12.11 is analogous to the proof of Theorem 12.10.

Example 12.12. For $p \neq 2$, let

$$p^* = (-1)^{\frac{p-1}{2}}p.$$

Moreover, let 2^* denote one of the numbers -4, 8, -8. Then every quadratic number field can be written uniquely in the form $k = \mathbb{Q}(\sqrt{p_1^* \cdots p_n^*})$, where $p_1, \ldots, p_n$ denote arbitrary pairwise distinct primes.

For $a \in \mathbb{Z}$ and an odd prime p, we define $[a,p]$ by

$$[a,p] = 1 \quad \text{for } p \nmid a,\ \left(\frac{a}{p}\right) = -1,$$
$$[a,p] = 0 \quad \text{otherwise}.$$

Moreover we put

$$[p^*,2] = 1 \quad \text{for } p^* \equiv 5 \bmod 8,$$
$$[p^*,2] = 0 \quad \text{otherwise}.$$

Let $K = \mathbb{Q}(\sqrt{p_1^* \cdots p_n^*})$ be an imaginary quadratic number field, and L/K the maximal unramified 2-extension. Then $\mathrm{Gal}(L/K)$ has a minimal presentation

$$1 \longrightarrow R_3 \longrightarrow H_1 \longrightarrow \mathrm{Gal}(L/K) \longrightarrow 1,$$

where H_1 is the free pro-p group with generators $w_2, \ldots, w_n$, and where R_3 is generated as a normal subgroup of H_1 by the relations $\rho_1, \ldots, \rho_n$ given by

$$\rho_1 \equiv \prod_{\nu=2}^{n} w_\nu^{2[p_\nu^*, p_1]} \pmod{H_1^{(3,p)}},$$

$$\rho_\mu \equiv w_\mu^{2[p_1^*, p_\mu]} \prod_{\nu=2}^{n} (w_\nu^2 w_\mu^2 [w_\nu, w_\mu])^{[p_\nu^*, p_\mu]} \pmod{H_1^{(3,p)}}, \quad \mu = 2, \ldots, n.$$

Example 12.13. Let $K = \mathbb{Q}(\sqrt{p_1^* \cdots p_n^*})$ be a real quadratic number field, and let L/K be the maximal 2-extension unramified outside ∞. Then $\mathrm{Gal}(L/K)$ has the same presentation as $\mathrm{Gal}(L/K)$ in Example 12.12.

Example 12.14. Let $K = \mathbb{Q}(\sqrt{p_1^* p_2^* p_3^*})$, and let L be defined as in Examples 12.12 and 12.13. Then $\mathrm{Gal}(L/K)$ is isomorphic to Klein's four group if and only if

$$\left(\frac{p_i^*}{p_j}\right) = 1, \ \left(\frac{p_j^*}{p_i}\right) = \left(\frac{p_i^*}{p_l}\right) = \left(\frac{p_l^*}{p_j}\right) = -1$$

for some permutation ijl of the digits 123.

$\mathrm{Gal}(L/K)$ is isomorphic to the quaternion group if and only if

$$\left(\frac{p_2^* p_3^*}{p_1}\right) = \left(\frac{p_3^* p_1^*}{p_2}\right) = \left(\frac{p_1^* p_2^*}{p_3}\right) = 1$$

and

$$\left(\frac{p_i^*}{p_j}\right) = \left(\frac{p_j^*}{p_i}\right) = -1$$

for two different $i, j \in \{1, 2, 3\}$.

12.3 A Criterion for Infinite Class Field Towers

In the following, k is a finite extension of $\mathbb{Q}$ and K/k an arbitrary p-extension. We want to prove a theorem saying that the p-class field tower of K is infinite if the number of primes ramified in K/k is larger than a certain bound depending only on k:

Theorem 12.15. *Let $p \neq 2$, and let K/k be a finite p-extension in which u primes are ramified. Let m denote the p-rank of the unit group of k, i.e.,*

$$m = r(k) - 1 + \delta(k),$$

where $r(k)$ denotes the number of infinite primes of k.

If

$$u \geq \begin{cases} 2m+5 & \textit{if } p \geq 5 \\ 2m+6 & \textit{if } p = 3, \end{cases} \tag{12.16}$$

then the p-class field tower of K is infinite.

More exactly, it is sufficient that u satisfy the bound

$$u \geq m + a + \sqrt{a^2 + 2am}, \tag{12.17}$$

where

$$a = \frac{2^{p-1}}{2^{p-2}-1} \quad \textit{if } p \geq 5,$$
$$a = \frac{27}{10} \quad \textit{if } p = 3.$$

Proof. It follows from

$$u > 2m + 5$$

that

$$u \geq m + \tfrac{5}{2} + \sqrt{(\tfrac{5}{2})^2 + 2(\tfrac{5}{2})m + m^2} \geq m + \tfrac{5}{2} + \sqrt{(\tfrac{5}{2})^2 + 2(\tfrac{5}{2})m}.$$

Since

$$a = \frac{2^{p-1}}{2^{p-2}-1} \leq \frac{5}{2} \quad \text{for } p \geq 5,$$

this implies that for $p \geq 5$ the bound (12.17) is better than (12.16). Something similar holds for $p = 3$.

In the following we shall use the notation introduced at the beginning of this chapter. We have to show that

$$\mathrm{Gal}\,(K_\varnothing/k) \simeq G_S/\langle \mathcal{T}_\mathfrak{p} \cap H_L \,|\, \mathfrak{p} \in S\rangle_{G_S}$$

is infinite.

By Lemma 4.14 there exists a normal subgroup $H_\mathfrak{p}$ of the decomposition group $\mathcal{Z}_\mathfrak{p}$ containing $\mathcal{T}_\mathfrak{p}$ and with index p in $\mathcal{Z}_\mathfrak{p}$. It suffices to prove that

$$G = G_S/\langle H_\mathfrak{p} \,|\, \mathfrak{p} \in S\rangle_{G_S}$$

is infinite.

For $\mathfrak{p} \in S$ with $N(\mathfrak{p}) \equiv 1 \pmod{p}$, the group $H_\mathfrak{p}$ is generated by $\tau_\mathfrak{p}^p$. For $\mathfrak{p} \in S$ and $\mathfrak{p} \mid p$, $\mathcal{Z}_\mathfrak{p}$ is generated by elements

$$\sigma_\mathfrak{p}, \tau_{1\mathfrak{p}}, \ldots, \tau_{n_\mathfrak{p}\mathfrak{p}}, \quad n_\mathfrak{p} = (k_\mathfrak{p} : \mathbb{Q}_p) + \delta(k_\mathfrak{p}),$$

where $\tau_{1\mathfrak{p}}, \ldots, \tau_{n_\mathfrak{p}\mathfrak{p}} \in \mathcal{T}_\mathfrak{p}$ (see Section 11.4). Put $\tau_{1\mathfrak{p}} = \tau_\mathfrak{p}$. Then $\tau_{1\mathfrak{p}}, \ldots, \tau_{n_\mathfrak{p}\mathfrak{p}}$ can be chosen in such a way that $H_\mathfrak{p}$ is generated as a normal subgroup of $\mathcal{Z}_\mathfrak{p}$ by

$$\tau_\mathfrak{p}^p, \tau_{2\mathfrak{p}}, \ldots, \tau_{n_\mathfrak{p}\mathfrak{p}}, [\sigma_\mathfrak{p}, \tau_\mathfrak{p}].$$

Let

$$1 \longrightarrow R \longrightarrow F \longrightarrow G_S \longrightarrow 1$$

be a minimal presentation of G_S. By the remark to Theorem 11.3, R is generated as a normal subgroup of F by the local relations and at most $\dim B_S$ additional relations. Thus, according to the considerations in the preceding sections, we get a presentation

$$1 \longrightarrow R_1 \longrightarrow F \longrightarrow G \longrightarrow 1 \tag{12.18}$$

of G, where R_1 is generated by the relations

$$\begin{aligned} t_\mathfrak{p}^p, [s_\mathfrak{p}, t_\mathfrak{p}] \quad & \text{for } \mathfrak{p} \in S,\ N(\mathfrak{p}) \equiv 1 \pmod p, \\ t_\mathfrak{p}^p, [s_\mathfrak{p}, t_\mathfrak{p}], t_{2\mathfrak{p}}, \ldots, t_{n_\mathfrak{p}\mathfrak{p}} \quad & \text{for } \mathfrak{p} \in S,\ \mathfrak{p} \mid p, \end{aligned}$$

and at most $\dim \mathbf{B}_S$ additional relations.

Now we apply Theorem 7.20. The relations of the form $t_{\nu\mathfrak{p}}$ have level ≥ 1; $[s_\mathfrak{p}, t_\mathfrak{p}]$, as well as the at most $\dim \mathbf{B}_S$ 'unknown' relations have level ≥ 2, and by Theorem 7.12, $t_\mathfrak{p}^p$ has level $\geq p$. Thus in Theorem 7.20 we can put

$$r_1 = \sum_{\substack{\mathfrak{p} \in S \\ \mathfrak{p} \mid p}} (n_\mathfrak{p} - 1), \quad r_2 = u + \dim \mathbf{B}_S, \ r_p = u$$

and $r_\nu = 0$ for $\nu \geq 3$, $\nu \neq p$.

Moreover, by Theorem 11.8 we have

$$d = d(F) = d(G_S) = \sum_{\substack{\mathfrak{p} \in S \\ \mathfrak{p} \mid p}} (n_\mathfrak{p} - 1) - m + u + \dim \mathbf{B}_S.$$

Assume that G is finite. Then

$$\varphi(t) = 1 - (u + \dim \mathbf{B}_S - m)t + (u + \dim \mathbf{B}_S)t^2 + ut^p > 0$$

for $0 < t < 1$, and in particular we have

$$1 - (u - m)t + ut^2 + ut^p > 0$$

for $0 < t < 1$.

If we plug in the value

$$t = \frac{u - m}{2u},$$

then we find

$$1 - \frac{(u-m)^2}{4u} + \left(\frac{u-m}{u}\right)^{p-2} \frac{(u-m)^2}{2^p u} > 0. \tag{12.19}$$

On the other hand (12.17) implies that

$$(u-m-a)^2 \geq a^2 + 2am,$$

hence

$$(u-m)^2 \geq 2ua. \tag{12.20}$$

We now plug $a = 2^{p-1}/(2^{p-2}-1)$ into (12.20) and get after a little calculation

$$1 - \frac{(u-m)^2}{4u} + \frac{(u-m)^2}{u2^p} \leq 0.$$

Since

$$\frac{u-m}{u} \leq 1,$$

this contradicts (12.19). Thus G is infinite.

This proves Theorem 12.15 for $p \geq 5$. For $p = 3$ the claim follows by putting

$$t = \frac{u-m}{3u}$$

and using the same reasoning. □

The last part of the proof of Theorem 12.15 can also be applied to the case $p = 2$. For infinite primes we get $\mathcal{T}_\mathfrak{p} \cap H_K = \{1\}$, i.e., infinite primes do not yield any "additional" relations. Since k contains the 2nd roots of unity, we can omit one relation. If the number of infinite primes ramified in K/k is q, we therefore find as in the proof of Theorem 12.15: The 2-class field tower of K is infinite if

$$\varphi(t) = 1 - (u-m)t + (2u - z - z)t^2 \leq 0$$

for some t with $0 < t < 1$. Now $\varphi(t)$ attains its minimum for

$$t_0 = \frac{(u-m)^2}{2(2u-z-1)}.$$

We get

$$\varphi(t_0) = 1 - \frac{(u-m)^2}{4(2u-z-1)},$$

and $\varphi(t_0) \leq 0$ is equivalent to

$$u \geq m + 4 + \sqrt{16 + 8m - 4(z+1)}. \tag{12.21}$$

Theorem 12.16. *If K/k is a finite 2-extension, then the 2-class field tower of k is infinite if* (12.21) *holds.*

Example 12.17. In the case of complex quadratic number fields K, Theorem 12.16 does not give the best possible result. In this case, six ramified primes in $K/\mathbb{Q}$ suffice for making the 2-class field tower of K infinite.

Example 12.18. By making all the symbols $[p_\nu^*, p_\mu]$ in Example 12.12 vanish, one gets relations of level ≥ 3. By Theorem 7.21 there are therefore infinitely many complex quadratic number fields $K/\mathbb{Q}$ with four ramified primes and infinite 2-class field tower.

13. The Cohomological Dimension of G_S

In this chapter k is an arbitrary global field of finite type. We study the cohomology groups $\mathrm{H}^\nu(G_S)$ in more detail and prove in particular that $\mathrm{H}^3(G_S)$ vanishes under certain conditions.

In the following, S will always denote a set of primes of k satisfying the conditions of Section 11.2 and containing all infinite real primes if $p = 2$. For a finite extension K/k, let $D_S(K)$ denote the group of all divisors of K coprime to S, and $E_S(K)$ the S-unit group of K, that is, the group of all $\alpha \in K$ in whose prime ideal factorization only primes of S occur. For an infinite extension K/k, we put

$$D_S(K) = \varinjlim D_S(K_\nu), \quad E_S(K) = \varinjlim E_S(K_\nu),$$

the inductive limit being over all finite subextensions K_ν/k of K/k. In particular, we put

$$D_S(k_S) = D_S, \quad E_S(k_S) = E_S.$$

13.1 Cohomology of S-Unit Groups

We define a homomorphism $\varphi : k_S^\times/E_S \longrightarrow D_S$ by mapping αE_S for $\alpha \in k_S^\times$ to the component of the principal ideal (α) that is coprime to S. Obviously, φ is injective. In the corresponding exact sequence

$$0 \longrightarrow k_S^\times/E_S \longrightarrow D_S \longrightarrow B_S \longrightarrow 0, \tag{13.1}$$

B_S is a torsion group whose elements have order coprime to p. For an algebraic number field, this is proved as follows: If $\mathfrak{a} \in D_S$, and if $\mathfrak{a}$ lies in a finite extension K of k, then there is a κ coprime to p such that $\mathfrak{a}^\kappa$ is principal in the p-class field tower of K (Theorem 8.8); for function fields, one uses Theorem 8.11 instead.

It follows from Theorem 3.19 that $\mathrm{H}^\nu(G_S, B_S)$ vanishes for $\nu \geq 1$. By (13.1) we therefore have an isomorphism

$$\mathrm{H}^\nu(G_S, k_S^\times/E_S) \simeq \mathrm{H}^\nu(G_S, D_S)$$

for $\nu \geq 1$.

The cohomology groups $\mathrm{H}^\nu(G_S, D_S)$ can be computed easily. In general, we have

Theorem 13.1. *Let K/k be a normal extension unramified outside of S, and put $G = \mathrm{Gal}(K/k)$. Then there is an isomorphism*

$$\gamma_S : \mathrm{H}^\nu(G, D_S(K)) \longrightarrow \coprod_{\mathfrak{p} \notin S} \mathrm{H}^\nu(\mathcal{Z}_{\mathfrak{P}}, \mathbb{Z}), \tag{13.2}$$

where $\mathfrak{P}$ denotes a fixed prime divisor of $\mathfrak{p}$ in K, and where $\mathcal{Z}_{\mathfrak{P}}$ acts trivially on $\mathbb{Z}$.

Proof. Assume first that K/k is finite. Then we have the decomposition of G-modules

$$D_S(k) = \coprod_{\mathfrak{p} \notin S} D_{\mathfrak{p}},$$

where $D_{\mathfrak{p}}$ is the group of all divisors generated by the primes dividing $\mathfrak{p}$. Moreover, $D_{\mathfrak{p}} \simeq M_G^{\mathcal{Z}_{\mathfrak{P}}}(\mathbb{Z})$ as G-modules: an isomorphism is provided by the map that sends

$$\mathfrak{a} = \prod_{\mathfrak{Q} | \mathfrak{p}} \mathfrak{Q}^{v_{\mathfrak{Q}}(\mathfrak{a})}$$

to the function $f_{\mathfrak{a}} \in M_G^{\mathcal{Z}_{\mathfrak{P}}}(\mathbb{Z})$ with

$$f_{\mathfrak{a}}(g) = v_{g^{-1}\mathfrak{P}}(\mathfrak{a}) \quad \text{for } g \in G.$$

The claim now follows from Shapiro's Lemma (Theorem 3.9).

Since $\mathfrak{p} \notin S$ is unramified in K/k by assumption, the transfer to infinite extensions K/k poses no problems. □

We remark in passing that the isomorphism (13.2) is induced by the map $v_{\mathfrak{P}} : D_S \longrightarrow \mathbb{Z}$. The diagram

$$\begin{array}{ccccc} k_S^\times & \longrightarrow & k_S^\times / E_S & \longrightarrow & D_S \\ \downarrow & & & & \downarrow v_{\mathfrak{P}} \\ (k_S)_{\mathfrak{P}}^\times & & \xrightarrow{\quad v \quad} & & \mathbb{Z} \end{array} \tag{13.3}$$

in which v denotes the normalized addititive valuation with value group $\mathbb{Z}$, is commutative.

The map

$$E_S \longrightarrow k_S^\times \longrightarrow (k_S)_{\mathfrak{P}}^\times$$

induces a homomorphism

$$\psi : \mathrm{H}^2(G_S, E_S) \longrightarrow \coprod_{\mathfrak{p} \in S} \mathrm{H}^2(\mathcal{Z}_{\mathfrak{P}}, (k_S)_{\mathfrak{P}}^{\times}),$$

where $\mathfrak{P}$ as above is a fixed prime divisor of $\mathfrak{p}$ in k_S. Moreover, let

$$\chi : \coprod_{\mathfrak{p} \in S} \mathrm{H}^2(\mathcal{Z}_{\mathfrak{P}}, (k_S)_{\mathfrak{P}}^{\times}) \longrightarrow \mathbb{Q}/\mathbb{Z}$$

be the map assigning to each $(\alpha_{\mathfrak{p}})_{\mathfrak{p} \in S}$ the sum of the invariants of the $\alpha_{\mathfrak{p}}$ (cf. Section 8.9). Then we have

Theorem 13.2. *If S is a set of primes in k that contains all real places in case $p = 2$ and is not empty for $\chi(k) \neq 0$, then the sequence*

$$0 \longrightarrow \mathrm{H}^2(G_S, E_S) \xrightarrow{\ \psi\ } \textstyle\coprod_{\mathfrak{p} \in S} \mathrm{H}^2(\mathcal{Z}_{\mathfrak{P}}, (k_S)_{\mathfrak{P}}^{\times}) \xrightarrow{\ \chi\ } \mathbb{Q}/\mathbb{Z}$$

is exact.

Proof. From the sequence

$$0 \longrightarrow E_S \longrightarrow k_S^{\times} \longrightarrow k_S^{\times}/E_S \longrightarrow 0 \qquad (13.4)$$

we deduce the exactness of

$$\mathrm{H}^1(G_S, k_S^{\times}/E_S) \longrightarrow \mathrm{H}^2(G_S, E_S) \longrightarrow \mathrm{H}^2(G_S, k_S^{\times}) \longrightarrow \mathrm{H}^2(G_S, k_S^{\times}/E_S).$$

By Theorem 13.1 and the remark preceding it we have $\mathrm{H}^1(G_S, k_S^{\times}/E_S) = 0$. By Section 8.9, the morphism v from (13.3) induces an isomorphism

$$\mathrm{H}^2(\mathcal{Z}_{\mathfrak{P}}, (k_S)_{\mathfrak{P}}^{\times}) \longrightarrow \mathrm{H}^2(\mathcal{Z}_{\mathfrak{P}}, \mathbb{Z}).$$

Thus (13.3) induces the commutative diagram

$$\begin{array}{ccccccc}
 & & & & 0 & & \\
 & & & & \downarrow & & \\
0 & \longrightarrow & \mathrm{H}^2(G_S, E_S) & \longrightarrow & \mathrm{H}^2(G_S, k_S^{\times}) & \longrightarrow & \mathrm{H}^2(G_S, k_S^{\times}/E_S) \\
 & & & & \downarrow & & \downarrow{\scriptstyle \gamma_S} \\
 & & & & \coprod_{\mathfrak{p}} \mathrm{H}^2(\mathcal{Z}_{\mathfrak{P}}, (k_S)_{\mathfrak{P}}^{\times}) & \longrightarrow & \coprod_{\mathfrak{p} \notin S} \mathrm{H}^2(\mathcal{Z}_{\mathfrak{P}}, \mathbb{Z}) \\
 & & & & \downarrow & & \\
 & & & & \mathbb{Q}/\mathbb{Z} & &
\end{array}$$

where the first vertical sequence is (8.10) for $K = k_S$. By Hasse's Local-Global principle, this sequence is exact.

For completing the proof of Theorem 13.2 it only remains to remark that, by Section 8.9, the image of the homomorphism

$$\mathrm{H}^2(G_S, E_S) \longrightarrow \mathrm{H}^2(G_S, k_S^{\times}) \longrightarrow \textstyle\coprod_{\mathfrak{p} \notin S} \mathrm{H}^2(\mathcal{Z}_{\mathfrak{P}}, (k_S)_{\mathfrak{P}}^{\times})$$

vanishes, and that γ_S is an isomorphism by Theorem 13.1. □

Theorem 13.3. *If S is a set of primes in k that contains all real places in case $p = 2$ and is not empty for $\chi(k) \neq 0$, and when we denote by Cl_S the quotient group of the class group of k modulo the subgroup generated by the finite primes in S, then there is an isomorphism*

$$\mathrm{H}^1(G_S, E_S) \longrightarrow \mathrm{Cl}_S[p].$$

Proof. By (13.4) and Theorem 3.18, the sequence

$$k^\times \longrightarrow (k_S/E_S)^{G_S} \longrightarrow \mathrm{H}^1(G_S, E_S) \longrightarrow 0$$

is exact. From (13.1) we get the exact sequence

$$0 \longrightarrow (k_S^\times/E_S)^{G_S} \longrightarrow D_S(k) \longrightarrow B_0 \longrightarrow 0,$$

where B_0 is a p-primary torsion group. The kernel of the natural projection $D_S(k) \longrightarrow \mathrm{Cl}_S$ can be identified with $k^\times E_S/E_S$.

Collecting everything, we get the commutative and exact diagram

$$\begin{array}{ccccccccc}
 & & & & B_0 & & & & \\
 & & & & \uparrow & & & & \\
0 & \longrightarrow & k^\times E_S/E_S & \longrightarrow & D_S(k) & \longrightarrow & \mathrm{Cl}_S & \longrightarrow & 0 \\
 & & {\scriptstyle \mathrm{id}}\uparrow & & \uparrow & & \uparrow & & \\
0 & \longrightarrow & k^\times E_S/E_S & \longrightarrow & (k_S^\times/E_S)^{G_S} & \longrightarrow & \mathrm{H}^1(G_S, E_S) & \longrightarrow & 0 \\
 & & & & \uparrow & & \uparrow & & \\
 & & & & 0 & & 0 & &
\end{array}$$

where the map

$$\mathrm{H}^1(G_S, E_S) \longrightarrow \mathrm{Cl}_S \tag{13.5}$$

is induced by the other maps.

The map (13.5) is injective. The diagram therefore can be extended to the commutative and exact diagram in Figure 13.1 Since B_0 is a p-primary torsion group, so is C_0. This implies the claim. □

We now want to prove under certain conditions on S that $\mathrm{H}^3(G_S, E_S) = 0$; here S should satisfy one of the following conditions:

A. S is not empty; if $\chi(k) \neq 0$, it contains all prime divisors of $\mathfrak{p}$; if $p = 2$, it contains all real primes.

B. S contains almost all primes $\mathfrak{p}$ with $N(\mathfrak{p}) \equiv 1 \bmod p$; if $p = 2$, it contains all real primes.

Condition B is stronger than A only for number fields k.

We first prove the following

$$\begin{array}{ccccccccc}
 & & & & & & 0 & & \\
 & & & & & & \uparrow & & \\
 & & & & B_0 & \longrightarrow & C_0 & & \\
 & & & & \uparrow & & \uparrow & & \\
0 & \longrightarrow & k^\times E_S/E_S & \longrightarrow & D_S(k) & \longrightarrow & \mathrm{Cl}_S & \longrightarrow & 0 \\
 & & \uparrow & & \uparrow & & \uparrow & & \\
0 & \longrightarrow & k^\times E_S/E_S & \longrightarrow & (k_S^\times/E_S)^{G_S} & \longrightarrow & \mathrm{H}^1(G_S, E_S) & \longrightarrow & 0 \\
 & & & & \uparrow & & \uparrow & & \\
 & & & & 0 & & 0 & &
\end{array}$$

Fig. 13.1.

Lemma 13.4. *Let S be a set of primes in k satisfying condition A or B. Then the extension $(k_S)_{\mathfrak{P}}/k_{\mathfrak{p}}$ is infinite for all primes $\mathfrak{P} \mid \mathfrak{p}$.*

Proof. If $\chi(k) = 0$ in case A, then for all $n \geq 1$, k_S contains the maximal subfield of $\mathbb{Q}(\zeta_{p^n})$ of p-power degree over $\mathbb{Q}$. If $\chi(k) = 0$ in case B, then, for almost all primes $q \equiv 1 \bmod \mathfrak{p}$, k_S contains the maximal p-extension of $\mathbb{Q}$ unramified outside q. The claim follows in both cases from the well known decomposition law in cyclotomic fields.

If $\chi(k) \neq 0$, then k_S contains the maximal p-extension of the finite field of constants of k, which already gives the desired infinite extension of the completion. □

Theorem 13.5. *If S is a set of primes of k satisfying condition A or B, then* $\mathrm{H}^3(G_S, E_S) = 0$.

Proof. We first prove that $\mathrm{H}^3(G_S, k_S^\times) = 0$. Let α be an element of $\mathrm{H}^3(G_S, k_S^\times)$ represented by $\alpha_L \in \mathrm{H}^3(\mathrm{Gal}(L/k), L^\times)$, where L/k is a finite normal subextension of k_S/k. By Lemma 13.4 and Section 8.10, there always exists a subextension K/k containing L/k such that α_L is in the kernel of the inflation

$$\mathrm{H}^3(\mathrm{Gal}(L/k), L^\times) \longrightarrow \mathrm{H}^3(\mathrm{Gal}(K/k), K^\times),$$

and this gives $\alpha = 0$.

The cohomology sequence for

$$1 \longrightarrow E_S \longrightarrow k_S^\times \longrightarrow k_S^\times/E_S \longrightarrow 1$$

gives the exact sequence

$$\mathrm{H}^2(G_S, k_S^\times) \xrightarrow{\ \gamma\ } \mathrm{H}^2(G_S, k_S^\times/E_S) \longrightarrow \mathrm{H}^3(G_S, E_S) \longrightarrow 0.$$

The map γ is surjective: to see this, consider as above the commutative diagram

$$\begin{array}{ccc} \mathrm{H}^2(G_S, k_S^\times) & \xrightarrow{\gamma} & \mathrm{H}^2(G_S, k_S^\times/E_S) \\ \downarrow & & \downarrow \\ \mathrm{H}^2(\mathcal{Z}_{\mathfrak{P}}, (k_S)_{\mathfrak{P}}^\times) & \longrightarrow & \coprod_{\mathfrak{p}\notin S} \mathrm{H}^2(\mathcal{Z}_{\mathfrak{P}}, \mathbb{Z}) \end{array} \tag{13.6}$$

corresponding to (13.3). Since $\mathrm{H}^2(\mathcal{Z}_{\mathfrak{P}}, (k_S)_{\mathfrak{P}}^\times) \simeq \mathbb{Q}_p/\mathbb{Z}_p$ by Lemma 13.4 and Section 8.9, the surjectivity of γ can be read off (13.6). This implies the claim. □

13.2 The Case $\delta(k) = 1$

We now assume that k contains the p-th roots of unity. We identify the group μ_p generated by the p-th roots of unity with $\mathbb{F}_p$. This is only possible, however, if the Galois group acts trivially on μ_p.

Now consider the exact Kummer sequence

$$1 \longrightarrow \mu_p \longrightarrow E_S \xrightarrow{p} E_S^p \longrightarrow 1, \tag{13.7}$$

where p denotes the map 'raising to the p-th power'.

If S contains all primes dividing p (this is no restriction for function fields), we have $E_S^p = E_S$ by the theory of cyclic p-extensions. For the following, it is sufficient that the cohomology groups $\mathrm{H}^\nu(G_S, E_S/E_S^p)$ vanish for $\nu \geq 1$.We have

Lemma 13.6. *Assume that S contains almost all primes of k. Then for all $\nu \geq 1$ we have*

$$\mathrm{H}^\nu(G_S, E_S/E_S^p) = 0.$$

Proof. We may assume that k is a number field.

Let K/k be a finite normal extension, $\mathfrak{P}$ a prime in K not ramified in K/k and dividing p, and let $\mathfrak{p}$ be the prime in k below $\mathfrak{P}$. Let $E_{\mathfrak{P}}$ denote the unit group of $K_{\mathfrak{P}}$, and $F_{\mathfrak{P}}$ its subgroup of primary elements, that is, of all $\alpha \in K_{\mathfrak{P}}$ for which $K_{\mathfrak{P}}(\sqrt[p]{\alpha})/K_{\mathfrak{P}}$ is unramified.

Next let f be the inertia degree and e the ramification index of $K_{\mathfrak{P}}/\mathbb{Q}_p$. Since $K_{\mathfrak{P}}/k_{\mathfrak{p}}$ is unramified, any uniformizer π of $k_{\mathfrak{p}}$ is also a uniformizer of $K_{\mathfrak{P}}$. We can choose a normal basis of the residue class field of $K_{\mathfrak{P}}$ over $\mathbb{F}_p$ with representatives $\omega_1, \ldots, \omega_f$ in the ring of integers of $K_{\mathfrak{P}}$; here 'normal' means that automorphisms of $K_{\mathfrak{P}}/k_{\mathfrak{p}}$ permute the basis. By Hensel's structure theorem for $E_{\mathfrak{P}}$ (cf. H. Hasse [22, p. 229]), $E_{\mathfrak{P}}/E_{\mathfrak{P}}^p F_{\mathfrak{P}}$ is a vector space over $\mathbb{F}_p$ with basis

$$\left\{\overline{1 + \omega_i \pi^j} \mid i = 1, \ldots, f,\ j = 1, \ldots, \frac{pe}{p-1},\ j \not\equiv 0 \pmod{p}\right\},$$

and from the choice of $\omega_1, \ldots, \omega_f$ it is clear that $E_{\mathfrak{P}}/E_{\mathfrak{P}}^p F_{\mathfrak{P}}$ is free as a $\mathrm{Gal}\,(K_{\mathfrak{P}}/k_{\mathfrak{p}})$-module. Moreover it can be seen easily that the $\mathrm{Gal}\,(K/k)$-module

$$\mathcal{E}_{\mathfrak{p}} = \coprod_{\mathfrak{P}|\mathfrak{p}} E_{\mathfrak{P}}/E_{\mathfrak{P}}^p F_{\mathfrak{p}}$$

is induced, i.e.,

$$\mathrm{H}^\nu(\mathrm{Gal}\,(K/k), \mathcal{E}_{\mathfrak{p}}) = 0 \quad \text{for } \nu \geq 1. \tag{13.8}$$

We now put

$$\mathcal{E}(K) = \coprod_{\mathfrak{p}} \mathcal{E}_{\mathfrak{p}},$$

where the sum is over all $\mathfrak{p} \mid p$, $\mathfrak{p} \notin S$, and form the inductive limit

$$\mathcal{E} = \varinjlim \mathcal{E}(K)$$

over all finite subextensions K/k of k_S/k. By (13.6), we have

$$\mathrm{H}^\nu(G_S, \mathcal{E}) = 0 \quad \text{for } \nu \geq 1. \tag{13.9}$$

Now we construct a morphism $\varphi : E_S/E_S^p \to \mathcal{E}$. For $\alpha \in E_S$ let K/k be a finite subextension of k_S/k such that $\alpha \in K$. For a prime divisor $\mathfrak{P}$ in K with $\mathfrak{P} \mid \mathfrak{p} \mid p$ and $\mathfrak{p} \notin S$, we have $\alpha \in E_{\mathfrak{P}}$. The map

$$\alpha \longmapsto \alpha E_{\mathfrak{P}}^p F_{\mathfrak{P}}$$

therefore induces a morphism

$$\varphi_K : K \cap E_S \longrightarrow \mathcal{E}(K).$$

Passing to the inductive limit yields a morphism

$$\varphi' : E_S \longrightarrow \mathcal{E}.$$

By construction, its kernel consists of all $\alpha \in E_S$ with the following property: if K/k is a finite extension with $\alpha \in K$, then $K(\sqrt[p]{\alpha})/K$ is unramified for $\mathfrak{P} \mid \mathfrak{p} \notin S$. Thus $\sqrt[p]{\alpha} \in E_S$, hence $\alpha \in E_S^p$. The converse

$$E_S^p \subseteq \ker \varphi'$$

is obvious. This gives us a monomorphism

$$\varphi : E_S/E_S^p \longrightarrow \mathcal{E}.$$

Under the assumptions of Lemma 13.6, φ is surjective: in fact, let

$$\sum_{\mathfrak{P}} \alpha_{\mathfrak{P}} E_{\mathfrak{P}}^p F_{\mathfrak{P}}$$

be a representative of an element $\overline{\beta}$ of $\mathcal{E}$ in $\mathcal{E}(K)$. The Chinese Remainder Theorem implies the existence of an integral $\alpha \in K$ with

$$\alpha \equiv \alpha_{\mathfrak{P}} \pmod{\mathfrak{P}^{e'}}, \quad e' = \frac{pe}{p-1}$$

for the finitely many $\mathfrak{P} \mid \mathfrak{p}$, $\mathfrak{p} \notin S$. Then $\alpha \in E_S$, and the class of α gets mapped to $\overline{\beta}$ under φ.

Thus φ is an isomorphism. The claim now follows from (13.9). □

We now can prove the following theorem.

Theorem 13.7. *Assume that $\delta(k) = 1$ and let S be a set of primes in k satisfying condition A or B. If $p = 2$ and $\chi(k) = 0$, assume that k is totally complex. Then* cd $G_S \le 2$.

Proof. We have to show that $\mathrm{H}^2(G_S) = 0$. From (13.7) and Theorem 13.5 we get that the sequence

$$\mathrm{H}^2(G_S, E_S) \longrightarrow \mathrm{H}^2(G_S, E_S^p) \longrightarrow \mathrm{H}^3(G_S) \longrightarrow 0 \qquad (13.10)$$

is exact. Since by Lemma 13.6 the embedding $E_S^p \hookrightarrow E_S$ induces an isomorphism $\mathrm{H}^2(G_S, E_S^p) \longrightarrow \mathrm{H}^2(G_S, E_S)$, we can replace (13.10) by the isomorphism

$$\mathrm{H}^2(G_S, E_S)/p \longrightarrow \mathrm{H}^3(G_S).$$

It remains to show that $\mathrm{H}^2(G_S, E_S)/p = 0$. But this can be read off immediately from Theorem 13.2 by observing that all $\mathrm{H}^2(\mathcal{Z}_{\mathfrak{P}}, (k_S)_{\mathfrak{P}}^{\times})$ are isomorphic to $\mathbb{Q}_p/\mathbb{Z}_p$. □

The structure of $\mathrm{H}^2(G_S)$ is described by the following theorem.

Theorem 13.8. *If $\delta(k) = 1$ and if the set S of primes in k satisfies condition A or B, then the injective map $\text{Ш}_S \longrightarrow \text{Б}_S$ constructed in Section 11.2 is surjective, and there exists a map*

$$\coprod_{\mathfrak{p} \in S} \mathrm{H}^2(G_{\mathfrak{p}}) \xrightarrow{\text{inv}} \mathbb{F}_p$$

such that the sequence

$$0 \longrightarrow \text{Б}_S \longrightarrow \mathrm{H}^2(G_S) \longrightarrow \coprod_{\mathfrak{p} \in S} \mathrm{H}^2(G_{\mathfrak{p}}) \longrightarrow \mathbb{F}_p \longrightarrow 0$$

is exact.

Proof. In case A, (13.7) gives us the exact sequence

$$0 \longrightarrow \mathrm{H}^1(G_S, E_S)/p \longrightarrow \mathrm{H}^2(G_S) \longrightarrow \mathrm{H}^2(G_S, E_S)[p] \longrightarrow 0.$$

In case B, we know that

$$\mathrm{H}^2(G_S) \longrightarrow \mathrm{H}^2(G_S, E_S)[p] \longrightarrow 0 \qquad (13.11)$$

is exact, and since $\text{Б}_S = 0$ by Theorem 11.3, we get $\text{Ш}_S = 0$. □

From Theorem 13.2 we see that the sequence

$$\mathrm{H}^2(G_S, E_S)[p] \longrightarrow \coprod_{\mathfrak{p}\in S} \mathrm{H}^2(\mathcal{Z}_{\mathfrak{P}}, (k_S)^\times_{\mathfrak{P}})[p] \longrightarrow \mathbb{Q}/\mathbb{Z} \qquad (13.12)$$

is exact. As in Section 11.1 we have canonical isomorphisms

$$\mathrm{H}^2(\mathcal{Z}_{\mathfrak{P}}, (k_S)^\times_{\mathfrak{P}})[p] \xrightarrow{\sim} \mathrm{H}^2(G_{\mathfrak{p}}, k_{\mathfrak{p}}^\times)[p] \xrightarrow{\sim} \mathrm{H}^2(G_{\mathfrak{p}}), \qquad (13.13)$$

where the composite map

$$\begin{aligned}\mathrm{H}^2(G_S) &\longrightarrow \mathrm{H}^2(G_S, E_S)[p] \longrightarrow \mathrm{H}^2(\mathcal{Z}_{\mathfrak{P}}, (k_S)^\times_{\mathfrak{P}})[p] \\ &\longrightarrow \mathrm{H}^2(G_{\mathfrak{p}}, k_{\mathfrak{p}}^\times)[p] \longrightarrow \mathrm{H}^2(G_{\mathfrak{p}}, k_{\mathfrak{p}}^\times) \longrightarrow \mathrm{H}^2(G_{\mathfrak{p}})\end{aligned}$$

is the localization map.

With the help of (13.12) and (13.13) we get the desired map

$$\coprod_{\mathfrak{p}\in S} \mathrm{H}^2(G_{\mathfrak{p}}) \xrightarrow{\mathrm{inv}} \mathbb{F}_p \qquad (13.14)$$

using the canonical embedding

$$\mathbb{F}_p \hookrightarrow \mathbb{Q}/\mathbb{Z}.$$

Clearly, (13.14) is surjective, so together with (13.11) – (13.13) the exactness of

$$\mathrm{H}^2(G_S) \longrightarrow \coprod_{\mathfrak{p}\in S} \mathrm{H}^2(G_{\mathfrak{p}}) \longrightarrow \mathbb{F}_p \longrightarrow 0$$

follows. This proves Theorem 13.8 in case B. In case A, it only remains to remark that $\mathrm{H}^1(G_S, E_S)/p \simeq \mathrm{Cl}_S/p$ by Theorem 13.3 and $\mathrm{Cl}_S/p \simeq \mathbf{Б}_S$ by Theorem 12.3.

Remark. In case A, Theorem 13.8 implies for finite S that

$$\chi_2(G_S) = -r_2. \qquad (13.15)$$

In this case we can therefore prove Theorem 13.7 also using Theorem 11.9 and (13.15).

13.3 The Case $\delta(k) = 0$

We want to prove the following theorem.

Theorem 13.9. *If $\delta(k) = 0$ and S is a set of primes in k satisfying condition B, then* $\mathrm{cd}\, G_S \leq 2$.

The proof proceeds by reduction to the case $\delta(k) = 1$. We put $k(\zeta_p) = k'$ and let S' denote the set of all primes in k' dividing primes $\mathfrak{p}$ in k with $\mathfrak{p} \in S$ or $N(\mathfrak{p}) \not\equiv 1 \pmod p$. For a normal extension L/K we put

$$\mathrm{H}^\nu(\mathrm{Gal}\,(L/K), \mathbb{F}_p) = \mathrm{H}^\nu(L/K), \quad \nu = 0, 1, \ldots,$$

where $\mathrm{Gal}\,(L/K)$ acts trivially on $\mathbb{F}_p$.

We first prove two lemmas.

Lemma 13.10. *The assumptions being the same as in Theorem 13.9, we have*

$$\mathrm{H}^2(k'_{S'}/k_S k')^{\mathrm{Gal}\,(k_S k'/k_S)} = 0.$$

Proof. Let $K_1 \subseteq K_2$ subextensions of $k_S k'/k'$ that are finite and normal over k, and let S_1 and S_2 denote the set of all primes in K_1 and K_2 dividing $\mathfrak{p}$ in k with $\mathfrak{p} \in S$ or $N(\mathfrak{p}) \not\equiv 1 \pmod p$. Then $\text{Б}_{S_1} = \text{Б}_{S_2} = 0$, and by Theorem 13.8 we have an exact and commutative diagram

$$\begin{array}{ccccc} 0 & \longrightarrow & \mathrm{H}^2(k'_{S'}/K_1) & \longrightarrow & \coprod_{\mathfrak{p}_1 \in S_1} \mathrm{H}^2(G_{\mathfrak{p}_1}) \\ & & \downarrow & & \downarrow \\ 0 & \longrightarrow & \mathrm{H}^2(k'_{S'}/K_2) & \longrightarrow & \coprod_{\mathfrak{p}_2 \in S_2} \mathrm{H}^2(G_{\mathfrak{p}_2}). \end{array} \tag{13.16}$$

We now form the inductive limit over all finite normal subextensions K_1/k' of $k_S k'/k'$. By Lemma 13.4 and Theorem 9.3 we get

$$\varinjlim \coprod_{\mathfrak{p}_1 \in S_1} \mathrm{H}^2(G_{\mathfrak{p}_1}) = 0.$$

This implies the claim. □

Lemma 13.11. *The assumptions being as in Theorem 13.9, we have*

$$\mathrm{H}^\nu(k'_{S'}/k_S) = 0 \quad \textit{for } \nu = 1, 2.$$

Proof. The claim $\mathrm{H}^1(k'_{S'}/k_S) = 0$ is proved in the same way as the claim $\mathrm{H}^1(\widehat{k'}/\widehat{k}) = 0$ in the proof of Theorem 9.6. Next,

$$\mathrm{H}^2(k'_{S'}/k_S) \simeq \mathrm{H}^2(k'_{S'}/k_S k')^{\mathrm{Gal}\,(k_S k'/k_S)}$$

by Theorem 3.15, hence this group vanishes by Lemma 13.10. □

Proof of Theorem 13.9. By Lemma 13.11 and Theorem 3.14, the inflation

$$\mathrm{H}^3(k_S/k) \longrightarrow \mathrm{H}^3(k'_{S'}/k)$$

is injective. On the other hand, $\mathrm{H}^3(k'_{S'}/k) \simeq \mathrm{H}^3(k'_{S'}/k')^{\mathrm{Gal}\,(k'/k)}$ by Theorem 3.15, hence it vanishes by Theorem 13.7. □

Now we want to prove an analog of Theorem 13.8. To this end, we first have to consider the sequence

$$0 \longrightarrow \mathrm{H}^2(k'_{S'}/k') \longrightarrow \coprod_{\mathfrak{p}' \in S'} \mathrm{H}^2(G_{\mathfrak{p}'}) \longrightarrow \mathbb{F}_p \longrightarrow 0 \tag{13.17}$$

as a sequence of $\mathrm{Gal}\,(k'/k)$-modules. Let us put $H = \mathrm{Gal}\,(k'/k)$.

We first give $\coprod_{\mathfrak{p}'\in S'} \mathrm{H}^2(G_{\mathfrak{p}'})$ an H-module structure that is compatible with the localization map $\varphi_{S'}^*$. For $s\in H$, let $\overline{s}$ be a lift of s to $k'_{S'}$ that leaves k_S fixed. Moreover, for $\mathfrak{p}\in S$ let $\mathfrak{p}'$ and $\mathfrak{P}'$ be extensions of $\mathfrak{p}$ to k' and $k'_{S'}$, respectively, such that $\mathfrak{P}'\mid\mathfrak{p}'$. Then $\overline{s}$ induces an isomorphism

$$(k'_{S'})_{\mathfrak{P}'} \longrightarrow (k'_{S'})_{\overline{s}\mathfrak{P}'},$$

which in turn induces an isomorphism

$$\mathrm{H}^2((k'_{S'})_{\mathfrak{P}'}/k'_{\mathfrak{p}'}) \xrightarrow{\ \widetilde{s}\ } \mathrm{H}^2((k'_{S'})_{\overline{s}\mathfrak{P}'}/k'_{s\mathfrak{p}'}).$$

such that the diagram

$$\begin{array}{ccc} & \mathrm{H}^2((k_S)_{\mathfrak{P}}/k_{\mathfrak{p}}) & \\ \swarrow & & \searrow \\ \mathrm{H}^2((k'_{S'})_{\mathfrak{P}'}/k'_{\mathfrak{p}'}) & \longrightarrow & \mathrm{H}^2((k'_{S'})_{\overline{s}\mathfrak{P}'}/k'_{s\mathfrak{p}'}) \end{array} \tag{13.18}$$

commutes ($\mathfrak{P}$ denotes the restriction of $\mathfrak{P}'$ to k_S).

Now let $\varphi_{\mathfrak{p}'} : (k'_{S'})_{\mathfrak{P}'} \hookrightarrow \widehat{k}'_{\mathfrak{p}'}$ denote an embedding. By our arguments in the proof of Theorem 11.1 and Lemma 13.4, the induced map

$$\varphi_{\mathfrak{p}'}^* : \mathrm{H}^2((k'_{S'})_{\mathfrak{P}'}/k_{\mathfrak{p}'}) \longrightarrow \mathrm{H}^2(G'_{\mathfrak{p}'})$$

is an isomorphism. We define $\varphi_{s\mathfrak{p}'}$ and $\varphi_{s\mathfrak{p}'}^*$ correspondingly.

Now we define $s : \mathrm{H}^2(G_{\mathfrak{p}'}) \longrightarrow \mathrm{H}^2(G_{s\mathfrak{p}'})$ as the isomorphism making the diagram

$$\begin{array}{ccc} \mathrm{H}^2((k'_{S'})_{\mathfrak{P}'}/k'_{\mathfrak{p}'}) & \xrightarrow{\ \widetilde{s}\ } & \mathrm{H}^2((k'_{S'})_{\overline{s}\mathfrak{P}'}/k'_{s\mathfrak{p}'}) \\ {\scriptstyle\varphi_{\mathfrak{p}'}^*}\downarrow & & {\scriptstyle\varphi_{s\mathfrak{p}'}^*}\downarrow \\ \mathrm{H}^2(G_{\mathfrak{p}'}) & \xrightarrow[s]{} & \mathrm{H}^2(G_{s\mathfrak{p}'}) \end{array}$$

commutative.

Considerations analogous to those in Section 11.1 show that s does not depend on the choice of $\widetilde{s}$, $\mathfrak{P}'$, $\varphi_{\mathfrak{p}'}$ or $\varphi_{s\mathfrak{p}'}$.

Let $i_{\mathfrak{p}'} : \mathrm{H}^2(G_{\mathfrak{p}}) \longrightarrow \mathrm{H}^2(G_{\mathfrak{p}'})$ be the map induced by the embedding $\widehat{k}_{\mathfrak{p}} \hookrightarrow \widehat{k}_{\mathfrak{p}'}$. By (13.18) we have a commutative diagram

$$\begin{array}{ccc} & \mathrm{H}^2(G_{\mathfrak{p}}) & \\ {\scriptstyle i_{\mathfrak{p}'}}\swarrow & & \searrow{\scriptstyle i_{s\mathfrak{p}'}} \\ \mathrm{H}^2(G_{\mathfrak{p}'}) & \xrightarrow{\ s\ } & \mathrm{H}^2(G_{s\mathfrak{p}'}) \end{array} \tag{13.19}$$

The map

$$\psi_{\mathfrak{p}'} : \mathrm{H}^2(G_{\mathfrak{p}'}) \longrightarrow \mathbb{F}_p$$

occurring in (13.17) is defined as the composition of maps

$$\mathrm{H}^2(G_{\mathfrak{p}'}) \xrightarrow{\ \sim\ } \mathrm{H}^2(G_{\mathfrak{p}'}, \mu_p) \xrightarrow{\ \sim\ } \mathrm{H}^2(G_{\mathfrak{p}'}, \widehat{k}_{\mathfrak{p}'})[p] \xrightarrow{\mathrm{inv}_{k_{\mathfrak{p}'}}} \mathbb{Q}/\mathbb{Z} \longrightarrow \mathbb{F}_p .$$

By our considerations in the second proof of Theorem 10.5 and the transformation rule (iv) of $\mathrm{inv}_{k_{\mathfrak{p}'}}$ in Section 8.9 we get a commutative diagram

$$\begin{array}{ccc} \mathrm{H}^2(G_{\mathfrak{p}'}) & \xrightarrow{\ s\ } & \mathrm{H}^2(G_{s\mathfrak{p}'}) \\ \psi_{\mathfrak{p}'} \big\downarrow & & \big\downarrow \psi_{s\mathfrak{p}'} \\ \mathbb{F}_p & \xrightarrow[s]{} & \mathbb{F}_p \end{array} \tag{13.20}$$

where the action of s on $\mathbb{F}_p$ is given by

$$s\bar{a} = \bar{a}\overline{a(s)}^{-1},$$
$$s\zeta_p = \zeta_p^{a(s)}, \quad a, a(s) \in \mathbb{Z}.$$

After these preliminaries we now can prove the following theorem:

Theorem 13.12. *If $\delta(k) = 0$ and S is a set of primes of k satisfying condition B, then the localization map*

$$\varphi_S^* : \mathrm{H}^2(G_S) \longrightarrow \coprod_{\mathfrak{p} \in S} \mathrm{H}^2(G_\mathfrak{p})$$

is an isomorphism.

Proof. By Theorem 11.3 and $\text{Б}_S = 0$, the map φ_S^* is injective. We have a commutative diagram

$$\begin{array}{ccccc} \mathrm{H}^2(G_S) & \longrightarrow & \coprod_{\mathfrak{p}\in S} \mathrm{H}^2(G_\mathfrak{p}) & & \\ \big\downarrow & & \big\downarrow \gamma & & \\ \mathrm{H}^2(G_{S'}) & \longrightarrow & \coprod_{\mathfrak{p}'\in S'} \mathrm{H}^2(G_{\mathfrak{p}'}) & \xrightarrow[\mathrm{inv}]{} & \mathbb{F}_p \end{array}$$

Let $\alpha_\mathfrak{p} \in \mathrm{H}^2(G_\mathfrak{p})$, $\mathfrak{p} \in S$. It is sufficient to show that $\alpha_\mathfrak{p}$ is in the image of φ_S^*. By our general assumptions on sets S of primes in Section 11.2 we have $N(\mathfrak{p}) \equiv 1 \pmod p$. Thus $\mathfrak{p}$ splits completely in k'/k, and for a fixed prime divisor $\mathfrak{p}'$ of $\mathfrak{p}$ in k', $s\mathfrak{p}'$ runs through all prime divisors of $\mathfrak{p}$ in k' for $s \in H$.

By (13.19) we have

$$t\Big(\sum_{s\in H} i_{s\mathfrak{p}'}\alpha_\mathfrak{p}\Big) = \sum_{s\in H} i_{ts\mathfrak{p}'}\alpha_\mathfrak{p} = \sum_{s\in H} i_{s\mathfrak{p}'}\alpha_\mathfrak{p},$$

i.e.

$$t\gamma\alpha_{\mathfrak{p}} = \gamma\alpha_{\mathfrak{p}} \quad \text{for } t \in H, \tag{13.21}$$

and by (13.19) and (13.20), inv maps $\gamma\alpha_{\mathfrak{p}}$ to

$$\sum_{s \in H} \psi_{s\mathfrak{p}'} i_{s\mathfrak{p}'} \alpha_{\mathfrak{p}} = \sum_{s \in H} \psi_{s\mathfrak{p}'} s i_{\mathfrak{p}'} \alpha_{\mathfrak{p}} = \sum_{s \in H} \overline{a(s)}^{-1} \psi_{\mathfrak{p}'} i_{\mathfrak{p}'} \alpha_{\mathfrak{p}}. \tag{13.22}$$

As is easily seen, we have

$$\sum_{s \in H} \overline{a(s)}^{-1} = 0. \tag{13.23}$$

It follows from (13.21) – (13.23) that there is an element in $\mathrm{H}^2(G_{S'})^H$ which $\varphi^*_{S'}$ maps to $\gamma\alpha_{\mathfrak{p}}$. Since $\mathrm{H}^2(G_S) \simeq \mathrm{H}^2(G_{S'})^H$, the claim follows. $\square$

References

Chapter 1 collects the basic properties of profinite groups. We use the language of category theory, about which the readers may inform themselves e.g. in S. MacLane [44, Ch. 1].

First glimpses of the Galois theory of infinite extensions as discussed in Chapter 2 can already be found in R. Dedekind [9]. Let us cite the following remark from page 288:

> ...hieraus folgt, daß der Inbegriff $\mathfrak{E}$ aller ε in gewissem Sinne eine stetige Mannigfaltigkeit bildet, worauf wir hier nicht weiter eingehen.[1]

W. Krull [40] has given the exact definition of the topology of the Galois group of an infinite normal extension in 1928. These groups became interesting in their own right, however, only through the work of K. Iwasawa [25, 26] and Y. Kawada [28], where for the first time a group theoretical classification of Galois groups of more complicated infinite extensions was initiated.

A complete discussion of the general Galois theory of infinite extensions that differs from the one given here can be found in N. Bourbaki [3], and K. Gruenberg [20] gives another concise presentation.

Chapter 3 gives a selfcontained presentation of the cohomology of profinite groups. Only the cohomology of complexes, which can be acquired easily in S. Lang [42, Ch. 4], was assumed to be known. In particular, we avoid the use of spectral sequences and utilize the method of dimension shifting whenever possible.

Chapters 4 to 6 contain mostly material from J.-P. Serre [53, Ch. 1]. The theory of free pro-p groups is presented completely using ideas from M. Lazard [43]. Theorem 5.5 is due to H. Koch [37], and Theorem 6.14 to I.R. Shafarevich (see K. Höchsmann [23]).

For Chapter 7, we used several sources. Theorem 7.7 can be found, except for the criteria (ii) and (iii), with a different proof in A. Brumer [7, Cor. 5.3], and Theorem 7.16 goes back to M. Lazard [43]. The proof of Lemma 7.19 is due to E.B. Vinberg [62]. The proof of Theorem 7.20 goes back essentially to E.S. Golod & I.R. Shafarevich [19]. While they used a result corresponding

[1] ...this implies that the collection $\mathfrak{E}$ of all ε form in a certain sense a continuous manifold, which we will not discuss any further here.

to Lemma 7.19 and worked in the ring of formal power series, we followed P. Roquette [49] and used real power series. The connection between the structure of relations and the cup product (7.8) was given by J.-P. Serre [52].

In Chapter 8, we have cited several theorems and results from algebraic number theory without proof. This chapter is intended to give readers without a background in class field theory an access to understanding the connections. The proofs of the cited theorems can be found in the following references:

Section	Reference
8.1–8.4	E. Weiss [63], J.W.S. Cassels & A. Fröhlich [8, Chap. I, II].
8.5	H. Hasse [22, § 15].
8.6, 8.9	E. Artin & J. Tate [1, Ch. 5–8], [8, Ch. 6–7], [46, Part II], [51, Ch. 11–14]
8.7	[1, Ch. 9]
8.8	[1, Ch. 13], H. Zassenhaus [65]
8.10	[1, Ch. 7]
8.11	is a presentation of [51, Ch. 14, § 2]

The results of Chapter 9 were, as far as the structure of the Brauer group $\mathrm{H}^2(\mathrm{Gal}(\widehat{k}/k), \widehat{k}^\times)$ are concerned, already known in the 1930's (cf. M. Deuring [13]). Theorem 9.1 is essentially due to I.R. Shafarevich [54], Theorem 9.3 to K. Iwasawa [26] and Y. Kawada [28]. Whereas these articles use the method of field embeddings, we have used the exact Kummer sequence, a method going back to J.P. Serre [53, Ch. II.2]. The reduction theorem 9.6 is due to K. Höchsmann [24].

Chapter 10: Theorems 10.1 and 10.2 are well known for finite extensions. The result on the relation rank in Theorem 10.3 is due to Y. Kawada [28]; other proofs were given by D.K. Faddeev & A.I. Skopin [14]

The proof presented here is due to H. Koch [37]. Theorem 10.5 was first proved by I.R. Shafarevich [54], also in a version for finite extensions. The idea in his proof is the same as the one used for our first proof. The second proof is due to K. Höchsmann [24]. More proofs were given by Y. Kawada [28] and J.P. Serre [53, Thm. II.3]. Theorem 10.9 is due to H. Koch [32]. The exact structure of G_S as a pro-p group was determined in the articles by A.I. Skopin [60], S.P. Demushkin [10, 11, 12], J.P. Serre [52] and J. Labute [41]. The last paper also contains an overview.

Chapter 11: Section 11.1 was taken from K. Höchsmann [23]. Theorems 11.1 and 11.2 were first proved in non-cohomological form by H. Koch [32]. Theorems 11.3 and 11.4 are published here for the first time. Theorem 11.5 was proved first by different methods by I.R. Shafarevich [58] and formed the departure point for the theory of the groups G_S. Section 11.3 contains standard material from class field theory. Section 11.4 leads to the results of the author [32, § 4], which were proved in the case of the rational numbers as a base field in a slightly different form by A. Fröhlich [15] (Examples 11.11, 11.12). Example 11.15 is due to J. Browkin [4], see also H. Koch [32, 33, 34]. Theorem 11.16 was proved in a weaker and non-invariant form in

H. Koch [32, Satz 6.1]. For $p = 2$, $k = \mathbb{Q}$, $S = \{2, \infty\}$, G_S was determined by G.N. Markshaitis [45]; this result can be read off easily from Example 11.18. More examples can be found in A. Brumer [6] and I.R. Shafarevich [58].

Chapter 12: Theorems 12.2 and 12.6 can be viewed as generalizations of Theorems 3 and 4 in A. Fröhlich [16] and [17], respectively. Example 12.5, in particular, is taken from [16]. The articles [16, 18] by A. Fröhlich contain more results on the structure of the p-component of absolutely abelian fields. Example 12.4 is due to K. Iwasawa [27] and was known, in the special case $p = 2$, already to H. Weber. Theorem 12.9 is a generalization of Theorem 1 in H. Koch [30]. A. Brumer [5] was the first to prove the theorem that the p-class field tower is infinite if there are sufficiently many primes in $\mathbb{Q}$ that are ramified in the base field K. Brumer's result for $p \neq 2$ follows easily from Theorem 12.15. See also P. Roquette [49].

Theorems 13.7 and 13.8 were proved by A. Brumer [6] for number fields satisfying condition A. For Theorem 13.12, see H. Koch [35] and K. Höchsmann [24], where the special case of the maximal p-extension is discussed.

Remark added in proof: L.W. Kuzmin has achieved farreaching results on the cohomological dimension of the groups G_S that are to appear in Izv. Akad. nauk SSSR.

Bibliography

1. E. Artin, J. Tate, *Class Field Theory*, Harvard 1961; 2nd ed. Addison-Wesley 1990 164

2. Z.I. Borevich, *On extensions without simple ramification of a regular local field* (Russian), Vestnik Leningrad. Univ. **11** (1956), no. 19, 41–47 2

3. N. Bourbaki, *Algébre*, Paris. Chap. V (Russian Transl. Moscow 1965) 163

4. J. Browkin, *On the generalized class field tower*, Bull. Acad. Polon. Sci. **11** (1963), 143–145 164

5. A. Brumer, *Ramification and class field towers of number fields*, Mich. Math. J. **12** (1965), 129–131 165

6. A. Brumer, *Galois groups of extensions of number fields with given ramification*, Mich. Math. J. **13** (1966), 33–40 165

7. A. Brumer, *Pseudocompact algebras, profinite groups and class formations*, J. Algebra **4** (1966), 442–470 163

8. J.W.S. Cassels, A. Fröhlich, *Algebraic number theory*, London 1967; reprint 1986 ix, 164, 168, 170

9. R. Dedekind, *Über die Permutationen des Körpers aller algebraischen Zahlen*, Ges. Werke **2**, Braunschweig 1931, p. 272–292 163, 175

10. S.P. Demushkin, *The group of a maximal p-extension of a local field* (Russian), Izv. Akad. Nauk SSSR Ser. Mat. **25** (1961), 329–346 164

11. S.P. Demushkin, *On 2-extensions of a local field* (Russian), Mat. Sibirsk Z. **4** (1963), 951–955 164

12. S.P. Demushkin, *Topological 2-groups with an even number of generators and a complete defining relation* (Russian), Izv. Akad. Nauk. SSSR **29** (1965), 3–10 164

13. M. Deuring, *Algebren*, Berlin 1935 164

14. D.K. Faddeev, A.I. Skopin, *Proof of a theorem of Kawada* (Russian), Dokl. Akad. Nauk SSSR **127** (1959), 529–530 164, 176

15. A. Fröhlich, *On fields of class two*, Proc. London Math. Soc. (3) **4** (1954), 235–256 164, 179

16. A. Fröhlich, *On the absolute class-group of Abelian fields*, J. London Math. Soc. **29** (1954), 211–217 2, 165

17. A. Fröhlich, *The generalization of a theorem of L. Rédei's*, Quart. J. Math. Oxford (2) **5** (1954), 130–140 2, 165

18. A. Fröhlich, *On the absolute class-group of Abelian fields (II)*, J. London Math. Soc. **30** (1955), 72–80 165

19. E.S. Golod, I.R. Shafarevich, *On the class field tower* (Russian), Izv. Akad. Nauk. SSSR **28** (1964), 261–272 163

20. K. Gruenberg, *Profinite Groups*, in: Cassels & Fröhlich [8, p. 116–127] 163

21. M. Hall, *The theory of groups*, New York 1959; reprint Chelsea 1976; Russ. transl. Moscow 1962 44, 67, 73, 74, 135

22. H. Hasse, *Zahlentheorie*, Akademie Verlag 1949; Engl. Transl.: *Number Theory*, Springer-Verlag 1980 154, 164

23. K. Höchsmann, *Über die Gruppe der maximalen l-Erweiterung eines globalen Körpers*, J. Reine Angew. Math. **219** (1965), 142–147 163, 164

24. K. Höchsmann, *l-extensions*, see Cassels & Fröhlich [8, p. 297–300] 164, 165

25. K. Iwasawa, *On solvable extensions of algebraic number fields*, Ann. Math. (II) **58** (1953), 548–572 2, 163

26. K. Iwasawa, *On Galois groups of local fields*, Trans. Amer. Math. Soc. **80** (1955), 448–469 163, 164

27. K. Iwasawa, *A note on class numbers of algebraic number fields*, Abh. Math. Sem. Univ. Hamburg **20** (1956), 257–258 165

28. Y.Kawada, *On the structure of the Galois group of some infinite extensions*, J. Fac. Sci. Univ. Tokyo **7** (1954), 1–18 163, 164, 176

29. H. Koch, *Über Darstellungsräume*, Math. Nachr. **26** (1963), 67–100 2

30. H. Koch, *Über den 2-Klassenkörperturm eines quadratischen Zahlkörpers*, J. Reine Angew. Math. **214/215** (1964), 201–206 165

31. H. Koch, *Über Galoissche Gruppen von $\mathfrak{p}$-adischen Zahlkörpern*, Math. Nachr. **29** (1965), 77–111 2

32. H. Koch, *l-Erweiterungen mit vorgegebenen Verzweigungsstellen*, J. Reine Angew. Math. **219** (1965), 30–61 2, 129, 164, 165

33. H. Koch, *l-Erweiterungen mit zwei Verzweigungsstellen*, Monatsber. Deutsch. Akad. Wiss. Berlin **7** (1965), 616–623. 164

34. H. Koch, *Über beschränkte Gruppen*, J. Algebra **3** (1966), 206–224 129, 164

35. H. Koch, *Über eine Vermutung von Höchsmann*, J. Reine Angew. Math. **225** (1967), 203–206 165

36. H. Koch, *Über die Hochschild-Serre-Sequenz*, Monatsber. Deutsch. Akad. Wiss. Berlin **8** (1966), 866–869

37. H. Koch, *Über die Dimension der Galoisschen Gruppen maximaler p-Erweiterungen*, Monatsber. Deutsch. Akad. Wiss. Berlin **10** (1968), 5–8 163, 164

38. H. Koch, *Zum Satz von Golod-Safarevic*, Math. Nachr. **42** (1969), 321–333

39. H. Koch, W. Zink, *Über die 2-Komponente der Klassengruppe quadratischer Zahlkörper*, Math. Nachr. **54** (1972), 309–333

40. W. Krull, *Galoissche Theorie der unendlichen algebraischen Erweiterungen*, Math. Ann. **100** (1928), 687–698 163

41. J. Labute, *Classification of Demushkin groups*, Can. J. Math. **19** (1966), 106–132 2, 164

42. S. Lang, *Algebra*, Reading 1965; Addison-Wesley 1984; Russ. Transl. Moscow 1968 163

43. M. Lazard, *Sur les groupes nilpotents et les anneaux de Lie*, Ann. Ec. Norm. Sup. **71** (1954), 101–190 163

44. S. MacLane, *Homology*, Springer Verlag 1963; reprint 1995; Russian Transl. Moscow 1966 163

45. G.N. Markschaitis, *On p-extensions with one critical number* (Russian), Izv. Akad. Nauk SSSR **27** (1963), 463–466 165

46. J. Neukirch, *Klassenkörpertheorie*, Bonner Math. Schriften **23** (1967); BI Mannheim 1969 ix, 164

47. L.S. Pontryagin, *Topological Groups* (Russian), Moscow; German Transl. Leipzig 1957; Engl. Transl. New York 1966 3, 7

48. H. Reichardt, *Konstruktion von Zahlkörpern mit gegebener Galoisgruppe von Primzahlpotenzordnung*, J. Reine Angew. Math. **177** (1937), 1–5 2

49. P. Roquette, *On class field towers*, see Cassels & Fröhlich [8, p. 231–249] 164, 165, 176

50. A. Scholz, *Konstruktion algebraischer Zahlkörper mit beliebiger Gruppe von Primzahlpotenzordnung*, Math. Z. **42** (1937), 161–188 2

51. J. P. Serre, *Corps Locaux*, Paris 1962; Engl. Transl. Springer-Verlag 1979 21, 164

52. J. P. Serre, *Structure de certain pro-p groupes*, Sém. Bourbaki 1962–63, exposé 252 164

53. J. P. Serre, *Cohomologie Galoisienne*, Lecture Notes in Math. **5**, Springer-Verlag 1964; Engl. Transl. Springer-Verlag 1997 163, 164, 180

54. I.R. Shafarevich, *On p-extensions* (Russian), Mat. Sbornik **20** (62) (1947), 351–361 164, 176

55. I.R. Shafarevich, *On the construction of fields with a given Galois group of order* l^α (Russian), Izv. Akad. Nauk SSSR, Ser. Mat. **18** (1954), 261–296 2

56. I.R. Shafarevich, *Construction of fields of algebraic numbers with given solvable Galois group* (Russian), Izv. Akad. Nauk SSSR. Ser. Mat. **18** (1954), 525–578 1

57. I.R. Shafarevich, *Algebraic number fields* (Russian), Proc. Congr. Stockholm 1962, p. 163–176; Engl. Transl.: Transl. Amer. Math. Soc. (II) **31** (1963), 25–39 1, 176

58. I.R. Shafarevich, *Extensions with prescribed ramification points* (Russian. French summary), IHES Publ. Math. **18** (1963), 71–95 2, 164, 165

59. I.R. Shafarevich, *Collected mathematical papers*, Translated from the Russian. Springer-Verlag 1989

60. A.I. Skopin, *p-extensions of a local field containing roots of unity of degree* p^m (Russian), Izv. Akad. Nauk SSSR **19** (1955), 445–470 164

61. A.I. Skopin, *Ideal of relations* (Russian), Trudy Mat. Inst. Steklova **80** (1964), 117–128 129

62. E.B. Vinberg, *On the theorem concerning the infinite-dimensionality of an associative algebra* (Russian), Izv. Akad. Nauk SSSR **29** (1965), 209–214 163, 176

63. E. Weiss, *Algebraic Number Theory*, New York 1963; reprint Dover 1998 164

64. A.V. Yakovlev, *The Galois group of the algebraic closure of a local field* (Russian), Izv. Akad. Nauk SSSR Ser. Mat. **32** (1968), 1283–1322 2

65. H. Zassenhaus, *Lehrbuch der Gruppentheorie*, Leipzig 1937; Engl. Transl. Chelsea 1949; 2nd ed. 1958; reprint Dover 1999 164

Notation

$\mathbb{N}$	set of natural numbers
$\mathbb{Z}$	ring of rational integers
$\mathbb{Q}$	field of rational numbers
$\mathbb{R}$	field of real numbers
$\mathbb{C}$	field of complex numbers
$\mathbb{Z}_p$	ring of p-adic integers
$\mathbb{Q}_p$	field of rational p-adic numbers
$[G, G]$	the (closed) commutator group of G
G^*	Pontryagin dual $G^* = \mathrm{Hom}\,(G, \mathbb{Q}/\mathbb{Z})$ of G
$\mathrm{Fr}(G)$	Frattini group $\overline{G^p[G,G]}$ of G
$\mathrm{cd}\ \ G$	cohomological dimension of G
$\chi(K)$	characteristic of the field K
$\chi(\mathfrak{p})$	characteristic of the residue class field mod $\mathfrak{p}$
K^+	additive group of the field K
$K^\times$	multiplicative group of the field K
$\mathbb{R}_+$	multiplicative group of positive real numbers
$\mathrm{Gal}\,(L/K)$	Galois group of the normal extension L/K
$D(K)$	group of divisors (fractional ideals) in K
$D_S(K)$	group of divisors in K coprime to S
$\mathrm{Cl}\,(K)$	ideal class group of K
$\mathrm{Cl}_S(K)$	S-class group of K
J_K	idele group of K
C_K	idele class group of K
$A(K)$	formation module of K
ζ_s	primitive s-th root of unity
μ_s	group of s-th roots of unity
p	prime number
K'	$K(\zeta_p)$
$\widehat{K}$	maximal p-extension of K
$\widehat{K'}$	maximal p-extension of K'
K_S	maximal extension unramified outside S
$\dim\ \ A$	$\dim A = \log_p \#A$ for finite p-groupS A

$\delta(k)$	For a fixed prime p we have $\delta(k) = 1$ or $= 0$ according as k contains the p-th roots of unity or not
A/q	A/qA for an abelian group A and $q \in \mathbb{N}$
$A[q]$	$\{a \in A \mid qa = 0\}$ for groups A and integers $q \in \mathbb{N}$
$A[p']$	elements of order coprime to p in a torsion group A
$\overline{a}$	the coset aN for $a \in A$ and some normal subgroup N of A
$\prod A_i$	the direct product of groups (modules, ...) A_i
$\coprod A_i$	the direct sum (coproduct) of groups (modules, ...) A_i.

Postscript

Helmut Koch, Franz Lemmermeyer

During the 30 years since the publication of the German original of this book, the theory of pro-p extensions was flourishing: new results were proved, new insights gained, and new conjectures emerged. We will take this opportunity to make a few comments on the origin and the history of the subject, and then briefly indicate some of the progress that has been made during the last few decades.

The Prehistory

It was Dedekind [9] who first investigated the Galois group of an infinite extension, namely of the field $\mathbb{Q}'$ of all p^nth roots of unity over the rationals $\mathbb{Q}$, apparently motivated by the fact that infinite extensions were 'off limits' for his contemporaries. In modern notation, he proved that $G = \mathrm{Gal}\,(\mathbb{Q}'/\mathbb{Q}) \simeq \mathbb{Z}_p^\times$, and that the Galois correspondence does not give a bijection between subgroups of G and subfields of $\mathbb{Q}'/\mathbb{Q}$.

Stiemke extended Dedekind's ideal theory to infinite extensions of $\mathbb{Q}$ in his dissertation [St], which was published posthumously by Emmy Noether. This article inspired Krull [Kr1], who was able to reestablish the Galois correspondence by introducing the topology on Galois groups that now bears his name, and by restricting the correspondence to closed subgroups of G. In the same year, Krull [Kr2, Kr3] also investigated ideal theory in infinite extensions of number fields.

Number theory in infinite extensions of $\mathbb{Q}$, in particular Hilbert's theory of ramification, was then studied by Herbrand [He1, He2], Scholz [Sl] and Moriya [Mo1], and class field theory was generalized to infinite extensions by Chevalley [Ch] (see also [Mo2]), but it seems that the time was not yet ripe for asking the right questions, and the whole topic must have looked like generalization for generalizations sake to most number theorists (comparable to the lukewarm reception of Hensel's p-adic numbers before Hasse proved the first local-global principle in 1920).

Golod-Shafarevich

The theory picked up pace after World War II, and the first result that should be mentioned was proved in 1947 by Shafarevich [54]. In modern notation it states that the Galois group G_k of the maximal p-extension of a $\mathfrak{p}$-adic number field k not containing the p-th roots of unity is a free pro-p group. In 1947, however, Shafarevich formulated this result as a theorem on certain finite p-extensions of k. The structure of the Galois group G_k was studied first by Kawada [28]. He showed that if k contains the p-th roots of unity, then G_k is a pro-p group with one generating relation. Faddeev and Skopin [14] gave a new proof for Kawada's result in 1959.

It seems that this article inspired I.R. Shafarevich to study the maximal p-extension k_S of a number field k that is unramified outside of S, with the goal in mind to prove that k_∞ is infinite for certain number fields k.

Here's the author's account of the early history of his collaboration with Shafarevich:

> In the spring of 1962, Shafarevich visited East Berlin. He asked me me to provide him with a copy of this article, and succeeded in transferring its main idea to global fields, thereby finding a bound for the number of relations of the group $G_S = \mathrm{Gal}(k_S/k)$. Later he told me that he first got the idea that in general there may be not enough relations to make G_S finite during a performance of the Rosenkavalier in the Berliner Staatsoper that we attended. Shafarevich presented his estimate of the relation rank at the ICM in Stockholm in 1962 in his survey [57] on algebraic number fields. In that lecture he posed the problem of finding a bound for the relation rank r in terms of the generator rank d of pro-p groups, with the hope that such a bound might prove the existence of infinite class field towers.
>
> Several mathematicians, in particular group theorists, must have tried to solve this problem; as Shafarevich told me later, Thompson was one of them. Nevertheless it was Shafarevich himself, in collaboration with his student Golod, who found a sufficiently strong bound. They showed in 1964 that a pro-p group is infinite if the relation rank r and the generator rank d satisfy the inequality $r \leq (d-1)^2$.
>
> The proof was quite complicated but was simplified almost immediately by Vinberg [62] and Gaschütz (cf. Roquette [49]), who also improved the bound to $r \leq \frac{1}{4}d^2$. The main idea in each of the three proofs is going from the group G to the complete group algebra $\mathbb{F}_p[[G]]$ and proving results on algebras.
>
> In the summer of 1962 I went to Verna in Bulgaria for holidays. During the hot afternoons I used to lay on the bed thinking about p-extensions. I had the idea that for maximal p-extensions of a number

field, all relations should come from the local relations; for restricted ramification one would have to factor out the inertia groups. Lacking stimulation, I did not pursue this any further. But in the autumn of 1962, when I learned about Shafarevich's estimate of the relation rank of G_S, I remembered these considerations and could develop the main ideas of my book in a very short time. A short version was published in the Monatsberichte der Deutschen Akademie der Wissenschaften 1963, and in the autumn of 1963 I gave talks on these results in Shafarevich's seminar in Moscow. When we met again in Oberwolfach in 1964, he had worked out a cohomological version of my results and showed me how the proofs could be formulated in an elegant way within the framework of the cohomology of pro-p groups. These ideas were taken up and refined by Höchsmann, who presented his investigations at the Brighton conference on algebraic number theory in 1965. This set the frame for the present book, which I wrote mainly while I was at the Steklov institute in Moscow during 1967-68.

Structure of G_S

Now let us mention some results on the structure of the groups G_S that have been obtained after the book had been published.

The results due to Kuzmin mentioned on page 165 have been published in [Ku1, Ku2]. He proved that G_S has cohomological dimension ≤ 2 if S contains all primes dividing p and all archimedean places, and if the base field k is totally complex if $p = 2$. Neumann [Ne1, Ne2] gave a simpler proof for this theorem and showed that the inequality (11.8) in Theorem 11.5 is actually an equality. Kuzmin [Ku3] also studied the question under which assumptions the maximal p-extension of $k_{\mathfrak{p}}$ for $\mathfrak{p} \in S$ is just a localization of k_S; again he assumes that S contains all primes dividing p as well as all archimedean primes. This condition is satisfied if $\mathfrak{p} \nmid p$.

In the case $\mathfrak{p}|p$, this condition is satisfied if the image of $G_{k,\mathfrak{p}}$ in G_S has infinite index; for this and related results (e.g. conditions under which G_S can be written as a free product of local groups) we refer to the monograph [NSW] by Neukirch, Schmidt & Wingberg. Let us also mention that Schmidt [Sc2] recently removed some of the problems caused by the prime 2 in connection with the structure of G_S when not all real places are in S.

Infinite Class Field Towers

The criterion of Golod-Shafarevich can be strengthened considerably in certain cases: Koch & Venkov [KV] showed that for odd primes p and complex quadratic base fields k, the condition $d(G_\varnothing) \geq 3$ implies that the p-class field tower of k is infinite.

This was generalized by Kisilevsky & Labute [KL] to general CM-fields (complex quadratic extensions of totally real number fields). Furuta [Fu] constructed fields of roots of unity with infinite p-class field towers; the criteria found by Schoof [Sch] led to much smaller examples, and later Wingberg [W1] came up with more general results that allowed him to prove that $\mathbb{Q}(\zeta_p)$ has infinite p class field tower for the prime $p = 157$ with irregularity index 2. The techniques of Schoof were taken up again by Maire [Ma2].

The result by Koch and Venkov remains true for the 'prime $p = 4$': if the class group of a complex quadratic number field has 4-rank ≥ 3, then its 2-class field tower is infinite (see Hajir [Ha2] for a simple proof); for similar results e.g. for cyclic cubic fields see Maire [Ma1].

Furuta [Fu] also showed that for every $s \geq 1$ there exist p-class field towers $K^{(n)}$ of cyclotomic fields K in which the p-class groups of every field $K^{(n)}$ have rank $\geq s$. This was improved by Hajir [Ha1], who showed that if a field K satisfies the Golod-Shafarevich criterion for infinite p-class field tower, then the rank of p-class groups of intermediate fields L are bounded below by a linear function of the degree $(L : K)$.

A general improvement of the Golod-Shafarevich inequality for p-groups seems is not possible for small values of the generator rank d: Newman and Wisliceny conjectured that for every prime p and every integer $d \geq 1$ there exists a finite pro-p group with d generators and $r = \lfloor d^2/4 \rfloor + 1$ relations, and together with Sauerbier they proved [NSaW] that this conjecture holds for $d \leq 6$.

It is, however, possible to improve Golod-Shafarevich by invoking invariants other than just r and d: Let $G_1 = \mathrm{Fr}(G) = [G,G]G^p$ (this is an elementary abelian group of order p^d), put $G_2 = [G_1,G]G_1^p$ or $G_2 = [G_1,G]G^p$ according as $p = 2$ or $p > 2$, and define e by $(G_1 : G_2) = p^e$. If G is a finite p-group, then Gaschütz & Newman [GN] proved

$$r > \frac{d^2}{2} + (-1)^p \frac{d}{2} + \Big(e - (-1)^p \frac{d}{2} - \frac{d^2}{4}\Big) \frac{d}{2}.$$

Finite Class Field Towers

Group-theoretical investigations on the possible Galois groups of finite p-class field towers of quadratic number fields began with the paper of Scholz & Taussky [ST], where 3-class field towers of complex quadratic number fields were investigated. Galois groups G of the Hilbert p-class field tower of complex quadratic fields (where p is an odd prime) have the following properties: G is a pro-p group with finite abelianization G/G', and there is an automorphism$\sigma \in \mathrm{Aut}(G)$ such that $1+\sigma$ acts trivially on G/G'. Groups with these properties are called Schur σ-groups; see Miyake [Mi] and Arrigoni [Ar]. Benjamin, Lemmermeyer & Snyder [BLS1, BLS2] have computed the Galois groups of 2-class field towers of length ≤ 2 for certain families of complex quadratic number fields. Bush [Bu] has recently found the first example of a quadratic number field whose 2-class field tower has length 3. Boston & Perry [BP] have computed the Galois groups of certain finite maximal extensions of $\mathbb{Q}$ unramified outside few primes, and their work was extended by Boston & Leedham-Green [BL].

Odlyzko's Discriminant Bounds

In the 1930s there was hope that improvements to Minkowski's discriminant bounds would lead to a proof that class field towers terminate; nowadays, class field towers are used to show that the best known discriminant bounds (due to Odlyzko) are close to best possible. The first constructions of infinite class field towers with this goal in mind were obtained by Martinet [Mt2, Mt3] and Schmithals [Sh]; Martinet's records were improved recently by Hajir & Maire [HM]. Martinet [Mt1] also showed how to use Odlyzko bounds to construct fields with class number 1 and large degree (up to 116).

Applications

Let p be a fixed prime, S be a finite set of places of a number field K containing all places above p, and let $G = G_S$ denote the Galois group of the maximal pro-p extension unramified outside of S. Define $G_1 = [G, G]$, ..., $G_{n+1} = [G, G_n]$.

The structure of G/G_1 is determined by class field theory, Fröhlich [15] determined the structure of G/G_2 for $K = \mathbb{Q}$, Ullom & Watt [UW] did the same for cyclotomic fields $\mathbb{Q}(\zeta_p)$ for regular primes p, and Movahhedi & Nguyen-Quang-Do [MT] for 'p-rational' number fields. Komatsu [Kom] studies this problem for real quadratic number fields K with class number not divisible by p and fundamental units ε with $\varepsilon^{p-1} \equiv 1 \bmod p^2$.

A particularly nice application of the results in this book was given by Bölling [Boe]: Callahan [Ca] had conjectured that $r_3(K) = r_3(k) - 1$, where $r_3(K)$ denotes the 3-rank of the class group of a non-normal cubic field K, and where $r_3(k)$ is the 3-rank of the class group of the quadratic subfield of the normal closure of $K/\mathbb{Q}$. This conjecture was later proved independently by Gerth and Gras. Bölling generalized this to extensions $K/\mathbb{Q}$ of odd prime degree p whose normal closure has degree $2p$ by proving the inequalities

$$r_p(k) - 1 \leq r_p(K) \leq (r_p(k) - 1)\frac{p-1}{2}.$$

Galois groups G of the maximal p-extension of $\mathbb{Q}$ unramified outside of two primes p and q have two generators and relation rank 1. The structure of G/G_2 is given in Thm. 11.10; Markshaitis [Mk1, Mk2] determined G/G_3 in the case $q \equiv 1 \bmod p^2$.

Fontaine-Mazur Conjecture

According to the Fontaine-Mazur Conjecture (see Fontaine & Mazur [FM]), G_S does not have infinite p-adic analytic quotients if S does not contain primes dividing p. In all cases where the infinitude of G_S can be proved by the Golod-Shafarevich inequality, one can also prove that G_S is not analytic. In fact, if the generator rank $d(G_S)$ of G_S is at least 2, the Golod-Shafarevich inequality $r(G_S) \leq \frac{1}{4}\, d(G_S)^2$ for the relation rank $r(G_S)$ implies that G_S is infinite and even non-analytic. Since analytic groups are characterized by the property of having finite rank (the open subgroups have bounded generator rank), the Fontaine-Mazur conjecture predicts that the p-rank of the class group of the layers of an infinite subquotient of k_S/k tends to infinity whenever S is away from p (see Hajir [Ha1]).

The Fontaine-Mazur conjecture can be restated as follows: every representation $\rho : G_S \longrightarrow \mathrm{GL}_n(\mathbb{Z}_p)$ has finite image. This was generalized by Boston to the conjecture that if R is a complete local Noetherian ring with finite residue field of characteristic p, then every representation $\rho : G_S \longrightarrow \mathrm{GL}_n(R)$ has finite image. To establish either conjecture, it suffices to replace G_S in the statement with an arbitrary just-infinite quotient of G_S (a group is called 'just-infinite' if it is infinite but has no proper infinite quotients).

While the Fontaine-Mazur conjecture remains out of reach at present, there are articles investigating some of its consequences; see Boston [Bo1, Bo2], Koch [Ko], Nomura [No1, No2, No3], and Wingberg [W2], as well as the contributions in du Sautoy, Segal, & Shalev [SSS].

Textbooks

Nowadays there are many textbooks discussing profinite groups from various points of view. For general topological groups there are Lutz [Lu] and Higgins [Hi]. Pontryagin duality for locally compact abelian groups is discussed by Morris [Mor].

For profinite groups we have the notes by Ribes [Ri] based on lectures by Neukirch, which have recently been reprinted; there are also new editions of Serre's Cohomologie Galoisienne [53]. The excellent report [Po] edited by Poitou is unfortunately out of print and would deserve more attention. The notes by Shatz [Sha] provide an introduction to the cohomology of profinite groups using spectral sequences; Fried & Jarden [FJ] also discuss profinite groups. Detailed introductions to profinite groups were given by Ribes & Zalesskii [RZ] as well as Wilson [Wil]. Dixon, du Sautoy Mann & Segal [DSMS] discuss analytic pro-p groups, and to complete the list, let us mention [NSW] once more. Finally Jarden's survey [Ja] gives a nice summary of results and open problems in infinite Galois theory.

Additional References

[Ar] M. Arrigoni, *On Schur σ-groups*, Math. Nachr. **192** (1998), 71–89 178

[BLS1] E. Benjamin, F. Lemmermeyer, C. Snyder, *Imaginary quadratic fields k with cyclic* $\mathrm{Cl}_2(k^1)$, J. Number Theory **67** (1997), no. 2, 229–245 178

[BLS2] E. Benjamin, F. Lemmermeyer, C. Snyder, *Imaginary quadratic fields k with* $\mathrm{Cl}_2(k) \simeq (2, 2^m)$ *and rank* $\mathrm{Cl}_2(k^1) = 2$, Pacific J. Math. 198 (2001), no. 1, 15–31 178

[Boe] R. Bölling, *On ranks of class groups of fields in dihedral extensions over $\mathbb{Q}$ with special reference to cubic fields*, Math. Nachr. **135** (1988), 275–310 179

[Bo1] N. Boston, *Some cases of the Fontaine-Mazur conjecture*, J. Number Theory **42** (1992), no. 3, 285–291 180

[Bo2] N. Boston, *Some cases of the Fontaine-Mazur conjecture. II*, J. Number Theory **75** (1999), no. 2, 161–169 180

[BL] N. Boston, Leedham-Green, *Explicit computation of Galois groups unramified at p*, J. Algebra, to appear 179

[BP] N. Boston, D. Perry, *Maximal 2-extensions with restricted ramification*, Journal of Algebra **232** (2000), 664–672 179

[Bu] M.R. Bush, *Computation of Galois groups associated to the 2-class towers of some quadratic fields*, preprint 2001 178

[Ca] T. Callahan, *The 3-class groups of non-Galois cubic fields. I, II*, Mathematika **21** (1974), 72–89; ibid. **21** (1974), 168–188 179

[Ch] C. Chevalley, *Généralisation de la théorie du corps de classes pour les extensions infinies*, J. Math. Pures Appl. **15** (1936), 359–371 175

[DSMS] J.D. Dixon, M.P.F. du Sautoy, A. Mann, D. Segal, *Analytic pro-p groups*, 2nd ed., Cambridge (1999) 180

[FM] J.-M. Fontaine, B. Mazur, *Geometric Galois representations*; in Elliptic curves, modular forms, and Fermat's last theorem (Hong Kong, 1993), 41–78, Ser. Number Theory, I, Cambridge MA, 1995. 180

[FJ] M.D. Fried, M. Jarden, *Field arithmetic*, Ergebnisse der Mathematik (3) 11, Springer-Verlag, Berlin, 1986 180

[Fu] Y. Furuta, *On class field towers amd the rank of ideal class groups*, Nagoya Math. J. **48** (1972), 147–157 178

[GN] W. Gaschütz, M.F. Newman, *On presentations of finite p-groups*, J. Reine Angew. Math. **245** (1970), 172–176 178

[Ha1] F. Hajir, *On the growth of p-class groups in p-class field towers*, J. Algebra **188** (1997), 256–271 178, 180

[Ha2] F. Hajir, *On a theorem of Koch*, Pacific J. Math. **176** (1996), no. 1, 15–18; Corr. ibid. **196** (2000), no. 2, 507–508 178

[HM] F. Hajir, C. Maire, *Tamely ramified towers and discriminant bounds for number fields*, Compositio Math. **128** (2001), no. 1, 35–53 179

[He1] J. Herbrand, *Théorie arithmétique des corps de nombres de degré infini. I*, Math. Ann. **106** (1932), 473–501 175

[He2] J. Herbrand, *Théorie arithmétique des corps de nombres de degré infini. II*, Math. Ann. **108** (1933), 699–717 175

[Hi] P.J. Higgins, *Topological Groups*, London Math. Soc. 1974 180

[Ja] M. Jarden, *Infinite Galois Theory*, in: Handbook of Algebra, Vol. 1, 269–319, Amsterdam 1996 180

[KL] H. Kisilevsky, J. Labute, *On a sufficient condition for the p-class field tower of a CM-field to be infinite*, Théorie des nombres, C. R. Conf. Int., Québec/Can. 1987, 556–560 (1989) 178

[Ko] H. Koch, *On maximal 2-extensions of $\mathbb{Q}$ with given ramification*, Preprint 2001 180

[KV] H. Koch, B.B. Venkov, *Über den p-Klassenkörperturm eines imaginär-quadratischen Zahlkörpers*, Soc. Math. France, Astérisque **24-25** (1975), 57–67 177

[Kom] K. Komatsu, *On the maximal p-extensions of real quadratic fields unramified outside p*, J. Algebra **123** (1989), 240–247 179

[Kr1] W. Krull, *Galoissche Theorie der unendlichen algebraischen Erweiterungen*, Math. Ann. **100** (1928), 687–698 175

[Kr2] W. Krull, *Idealtheorie in unendlichen algebraischen Zahlkörpern. I*, Math. Z. **29** (1928), 42–54 175

[Kr3] W. Krull, *Idealtheorie in unendlichen algebraischen Zahlkörpern. II*, Math. Z. **31** (1930), 527–557 175

[Ku1] L.V. Kuzmin, *Homology of profinite groups, Schur's multiplicator and class field theory* (Russian), Izv. Akad. Nauk SSSR, Ser. mat. **33** (1969), 1220–1254 177

[Ku2] L.V. Kuzmin, *The Tate module of an algebraic number field* (Russ.), Izv. Akad. Nauk SSSR, Ser. mat. **36** (1972), 267–327 177

[Ku3] L.V. Kuzmin, *Local extensions associated with l-extensions with given ramification* (Russian), Izv. Akad. Nauk SSSR, Ser. mat. **39** (1975); English translation in Math. USSR Izv. **9** (1975), 693–726 177

[Lu] D. Lutz, *Topologische Gruppen*, Mannheim 1976 180

[Ma1] C. Maire, *Tours de Hilbert des extensions cubiques cycliques de* $\mathbb{Q}$, Manuscripta Math. **92** (1997), 303–323 178

[Ma2] C. Maire, *Un raffinement du théorème de Golod-Safarevic*, Nagoya Math. J. **150** (1998), 1–11 178

[Mk1] G. Markshaitis, *Galois groups of p-extensions with two ramification points*, Lithuanian Math. J. **40** (2000), no. 1, 39–47 179

[Mk2] G. Markshaitis, *Construction of some p-extensions of the field of rational numbers*, Lithuanian Math. J. **40** (2000), no. 2, 140–147 179

[Mt1] J. Martinet, *Corps de nombre de classes* 1, Sémin. Théor. Nombres (1977–1978), Exp. No. 12, 8 pp., CNRS, Talence, 1978 179

[Mt2] J. Martinet, *Tours de corps de classes et estimations de discriminants*, Sémin. Théor. Nombres (1976–1977), Exp. No. 12, 4 pp., CNRS, Talence, 1977 179

[Mt3] J. Martinet, *Petits discriminants des corps de nombres*, Jour. arithmétiques, Exeter 1980, Lond. Math. Soc. Lect. Note Ser. **56** (1982), 151–193 179

[Mi] K. Miyake, *Some p-groups with two generators which satisfy certain conditions arising from arithmetic in imaginary quadratic fields*, Tohoku Math. J. (2) **44** (1992), no. 3, 443–469 178

[Mo1] M. Moriya, *Theorie der algebraischen Zahlkörper unendlichen Grades*, J. Fac. Sc. Hokkaido Imp. Univ. **3** (1935), 107–190; corr. ibid. **4** (1936), 121–122 175

[Mo2] M. Moriya, *Klassenkörpertheorie im Kleinen für die unendlichen algebraischen Zahlkörper*, J. Fac. Sci. Hokkaido Univ., Ser. I, Math. **5** (1936), 9–66 175

[Mor] S.A. Morris, *Pontryagin duality and the structure of locally compact abelian groups*, LMS Lecture Notes **29**, 1977 180

[MT] A. Movahhedi, T. Nguyen-Quang-Do, *Sur l'arithmétique des corps de nombres p-rationnels*, Sémin. Théor. Nombres Paris 1987–88, 155–200, Progr. Math., **81**, 1990. 179

[Ne1] O. Neumann, *On p-closed algebraic number fields with restricted ramification* (Russian), Izv. Ak. Nauk SSSR, Ser. mat. **39** (1975), 259–271 177

[Ne2] O. Neumann, *On p-closed number fields and an analogue of Riemann's existence theorem*, in: Algebraic Number Fields (A. Fröhlich, ed.), London (1977), 625–647 177

[NSaW] M.F. Newman, G. Sauerbier, J. Wisliceny, *Groups of prime-power order with a small number of relations*, Rostock. Math. Kolloq. **49** (1995), 141–154 178

[NSW] J. Neukirch, A. Schmidt, K. Wingberg, *Cohomology of Number Fields*, Springer Verlag Berlin 2000 vii, 177, 180

[No1] A. Nomura, *A remark on Boston's question concerning the existence of unramified p-extensions*, J. Number Theory **58** (1996), no. 1, 66–70 180

[No2] A. Nomura, *A remark on Boston's question concerning the existence of unramified p-extensions. II*, Proc. Japan Acad. Ser. A Math. Sci. **73** (1997), no. 1, 10–11 180

[No3] A. Nomura, *Embedding problems with restricted ramifications and the class number of Hilbert class fields*, Class field theory — its centenary and prospect (Tokyo, 1998), 79–86, Adv. Stud. Pure Math. **30**, Math. Soc. Japan, Tokyo, 2001 180

[Po] G. Poitou (ed.), *Cohomologie galoisienne des modules finis*, Travaux et Recherches Mathématiques, No. 13; Dunod, Paris 1967 180

[Ri] L. Ribes, *Introduction to profinite groups and Galois cohomology*, Queen's Papers in Pure and Applied Mathematics, 24 (1970); Corrected Reprint 1999 7, 180

[RZ] L. Ribes, P. Zalesskii, *Profinite groups*, Springer-Verlag, Berlin, 2000 180

[SSS] M. du Sautoy, D. Segal, A. Shalev (eds.), *New horizons in pro-p groups*, Birkhäuser Boston (2000) 180

[Sc1] A. Schmidt, *Extensions with restricted ramification and duality for arithmetic schemes*. Compos. Math. **100** (1996), 233–245

[Sc2] A. Schmidt, *On the relation between 2 and ∞ in Galois cohomology of number fields*, Compos. Math., to appear 177

[Sh] B. Schmithals, *Konstruktion imaginärquadratischer Körper mit unendlichem Klassenkörperturm*, Arch. Math. **34** (1980), no. 4, 307–312 179

[Sl] A. Scholz, *Zur Idealtheorie in unendlichen algebraischen Zahlkörpern*, J. Reine Angew. Math. **185** (1943), 113–126 175

[ST] A. Scholz, O. Taussky, *Die Hauptideale der kubischen Klassenkörper imaginärquadratischer Zahlkörper: ihre rechnerische Bestimmung und ihr Einfluß auf den Klassenkörperturm*, J. Reine Angew. Math. **171** (1934), 19–41 178

[Sch] R. Schoof, *Infinite class field towers of quadratic fields*, J. Reine Angew. Math. **372** (1986), 209–220. 178

[Sha] S.S. Shatz, *Profinite groups, artithmetic, and geometry*, Annals of Mathematics Studies, No. 67, 1972 180

[St] E. Stiemke, *Über unendliche algebraische Zahlkörper*, Math. Z. **25** (1926), 9–39 175

[UW] S.V. Ullom, S.B. Watt, *Generators and relations for certain class two Galois groups*, J. London Math. Soc. (2) **34** (1986), no. 2, 235–244 179

[Wil] J.S. Wilson, *Profinite Groups*, Oxford 1998 7, 180

[W1] K. Wingberg, *On the maximal unramified p-extension of an algebraic number field*, J. Reine Angew. Math. **440** (1993), 129–156. 178

[W2] K. Wingberg, *On the Fontaine-Mazur Conjecture for CM-Fields*, Compos. Math. (2002), to appear 180

Author Index

Subject Index

Springer Monographs in Mathematics

This series publishes advanced monographs giving well-written presentations of the "state-of-the-art" in fields of mathematical research that have acquired the maturity needed for such a treatment. They are sufficiently self-contained to be accessible to more than just the intimate specialists of the subject, and sufficiently comprehensive to remain valuable references for many years. Besides the current state of knowledge in its field, an SMM volume should also describe its relevance to and interaction with neighbouring fields of mathematics, and give pointers to future directions of research.

Abhyankar, S.S. **Resolution of Singularities of Embedded Algebraic Surfaces** 2nd enlarged ed. 1998
Andrievskii, V.V.; Blatt, H.-P. **Discrepancy of Signed Measures and Polynomial Approximation** 2002
Armitage, D.H.; Gardiner, S.J. **Classical Potential Theory** 2001
Arnold, L. **Random Dynamical Systems** 1998
Aubin, T. **Some Nonlinear Problems in Riemannian Geometry** 1998
Bang-Jensen, J.; Gutin, G. **Digraphs** 2001
Baues, H.-J. **Combinatorial Foundation of Homology and Homotopy** 1999
Brown, K.S. **Buildings** 3rd printing 2000 (1st ed. 1998)
Cherry, W.; Ye, Z. **Nevanlinna's Theory of Value Distribution** 2001
Ching, W.K. **Iterative Methods for Queuing and Manufacturing Systems** 2001
Crabb, M.C.; James, I.M. **Fibrewise Homotopy Theory** 1998
Dineen, S. **Complex Analysis on Infinite Dimensional Spaces** 1999
Elstrodt, J.; Grunewald, F. Mennicke, J. **Groups Acting on Hyperbolic Space** 1998
Fadell, E.R.; Husseini, S.Y. **Geometry and Topology of Configuration Spaces** 2001
Fedorov, Y.N.; Kozlov, V.V. **A Memoir on Integrable Systems** 2001
Flenner, H.; O'Carroll, L. Vogel, W. **Joins and Intersections** 1999
Gelfand, S.I.; Manin, Y.I. **Methods of Homological Algebra** 2nd ed. 2002
Griess, R.L.Jr. **Twelve Sporadic Groups** 1998
Ivrii, V. **Microlocal Analysis and Precise Spectral Asymptotics** 1998
Jorgenson, J.; Lang, S. **Spherical Inversion on SLn (R)** 2001
Kozlov, V.; Maz'ya, V. **Differential Equations with Operator Coefficients** 1999
Landsman, N.P. **Mathematical Topics between Classical & Quantum Mechanics** 1998
Lemmermeyer, F. **Reciprocity Laws: From Euler to Eisenstein** 2000
Malle, G.; Matzat, B.H. **Inverse Galois Theory** 1999
Mardesic, S. **Strong Shape and Homology** 2000
Narkiewicz, W. **The Development of Prime Number Theory** 2000
Parker, C.; Rowley, P. **Symplectic Amalgams** 2002
Prestel, A.; Delzell, C.N. **Positive Polynomials** 2001
Ranicki, A. **High-dimensional Knot Theory** 1998
Ribenboim, P. **The Theory of Classical Valuations** 1999
Rowe, E.G.P. **Geometrical Physics in Minkowski Spacetime** 2001
Rudyak, Y.B. **On Thom Spectra, Orientability and Cobordism** 1998
Ryan, R.A. **Introduction to Tensor Products of Banach Spaces** 2002
Saranen, J,; Vainikko, G. **Periodic Integral and Pseudodifferential Equations with Numerical Approximation** 2002
Schneider, P. **Nonarchimedean Functional Analysis** 2002
Serre, J-P. **Complex Semisimple Lie Algebras** 2001 (reprint of first ed. 1987)
Serre, J.-P. **Galois Cohomology** corr. 2nd printing 2002 (1st ed. 1997)
Serre, J-P. **Local Algebra** 2000
Springer, T.A.; Veldkamp, F.D. **Octonions, Jordan Algebras, and Exceptional Groups** 2000
Sznitman, A.-S. **Brownian Motion, Obstacles and Random Media** 1998
Tits, J.; Weiss, R.M. **Moufang Polygons** 2002
Uchiyama, A. **Hardy Spaces on the Euclidean Space** 2001
Üstünel, A.-S.; Zakai, M. **Transformation of Measure on Wiener Space** 2000
Yang, Y. **Solitons in Field Theory and Nonlinear Analysis** 2001

Druck: Strauss Offsetdruck, Mörlenbach
Verarbeitung: Schäffer, Grünstadt